Advances in
Insect Physiology

Volume 40

Advances in Insect Physiology

Spider Physiology and Behaviour: Physiology

edited by

Jérôme Casas

Institut de Recherche en Biologie de l'Insecte
IRBI UMR CNRS, Université de Tours, Tours, France

Volume 40

ELSEVIER

Amsterdam • Boston • Heidelberg • London • New York • Oxford
Paris • San Diego • San Francisco • Singapore • Sydney • Tokyo
Academic Press is an imprint of Elsevier

ACADEMIC PRESS

Academic Press is an imprint of Elsevier
32 Jamestown Road, London NW1 7BY, UK
Radarweg 29, PO Box 211, 1000 AE Amsterdam, The Netherlands
Linacre House, Jordan Hill, Oxford OX2 8DP, UK
225 Wyman Street, Waltham, MA 02451, USA
525 B Street, Suite 1900, San Diego, CA 92101-4495, USA

First edition 2011

ISBN: 978-0-12-387668-3
ISSN: 0065-2806

For information on all Academic Press publications
visit our website at elsevierdirect.com

Printed and bound by CPI Group (UK) Ltd, Croydon, CR0 4YY

Transferred to Digital Print 2011

Contents

Contributors

William G. Eberhard
Smithsonian Tropical Research Institute, Apartado, República de Panamá, USA; and Escuela de Biología, Universidad de Costa Rica, Ciudad Universitaria, Costa Rica

Sara L. Goodacre
School of Biology, University of Nottingham, Nottingham, United Kingdom

Lucia Kuhn-Nentwig
Institute of Ecology and Evolution, University of Bern, Bern, Switzerland

Wolfgang Nentwig
Institute of Ecology and Evolution, University of Bern, Bern, Switzerland

Reto Stöcklin
Atheris Laboratories, Geneva, Switzerland

William T. Wcislo
Smithsonian Tropical Research Institute, Apartado, República de Panamá, USA

Shawn M. Wilder
School of Biological Sciences, The University of Sydney, Sydney, New South Wales, Australia

Preface

The year 2011 is a record year for the number of books devoted to spiders: the third edition of the introduction *Biology of Spiders* by Rainer F. Foelix and an edited book by Marie E. Herberstein *Spider Behaviour* were both published in the first half of this year. The present two volumes of *Spider Physiology and Behaviour*, totaling over 550 pages, target a somewhat more specialized public and complement nicely the above works as well as the volumes edited in the 1980s by Wolfgang Nentwig (1987), *Ecophysiology of Spiders*, and Friedrich G. Barth (1985), *Spider Neurobiology*. The gap of over 20 years between these publications was a period of intensified research on this group of organisms, interspersed by the rare publication of books and monographs, in particular, F.G. Barth's *A Spider's World: Senses and Behaviour* (2002). A similar trend for several other groups of Arachnids can be observed.

The titles of these publications hint at one of the most fascinating aspect of spiders: their behaviour, its physiological basis, including the neurobiological components, and its consequences, including web construction. This fascination partly explains why many behavioural ecologists use spiders and why these arachnids became an accepted model of choice. These are relatively new trends of the past 20 years or so. Spider silk continued to raise sustained interest from a somewhat different group, in particular, from the material science quarters. The final chapter of the second volume *Spider Physiology and Behaviour: Behaviour* blends the two approaches of evolutionary biology and material sciences in the study of webs.

This rise of spiders as studied organisms and their secured place in the pool of accepted models is highly positive, not only in terms of visibility by the not-so-large community of scientists working with them but also in terms of attraction for students and of acceptance by the much larger community of ecologists: the role of spiders in nutrients flow and ecosystem services is large but under-appreciated. The planning of these two volumes identified other trends in need of attention. First, several chapters in the two edited books of the 1980s have no modern counterpart, not because the topics are out of fashion but because the expertise is lacking, worldwide, or because the field came to a full standstill. Second, the pool of tenured scientists working on these organisms is steady in comparison with other fast growing fields, with some variation between countries, if not continents. Thus, the increased visibility of spiders in main-stream journals is potentially the result of a mechanism by which a new generation of behavioural ecologists is replacing an older generation of physiologists and neuroethologists: more generic approaches to older problems, new journals and an increased level of assertiveness are among the hallmarks of this evolution. Finally, because behavioral ecologists make up a good portion of the most active community of scientists working on spiders, enhanced attention must be given to processes at the physiological and cellular levels. Indeed, most behavioural ecologists are poorly knowledgeable in these matters and may underestimate their interest and complexity. I hope that the volumes in this series, in particular, the first volume of this set, contribute to their education.

The splitting of a continuum from the cell to the organism is by definition arbitrary. A quick glance at the tables of contents shows that these two volumes constitute a set, not two independent books. Each chapter has been thoroughly checked by two external referees. I thank them and the authors for producing such a compelling body of knowledge.

Jérôme Casas

Venom Composition and Strategies in Spiders: Is Everything Possible?☆

Lucia Kuhn-Nentwig,* Reto Stöcklin,† and Wolfgang Nentwig*

*Institute of Ecology and Evolution, University of Bern, Bern, Switzerland
†Atheris Laboratories, Geneva, Switzerland

Abstract

This review on all spider venom components known by the end of 2010 bases on 1618 records for venom compounds from 174 spider species (=0.41% of all known species) belonging to 32 families (=29% of all existing spider families). Spiders investigated for venom research are either big (many mygalomorph species, Nephilidae, Ctenidae and Sparassidae) or medically important for humans (e.g. *Loxosceles* or *Latrodectus* species). Venom research widely ignored so far the two most species-rich families (Salticidae and

☆ Dedicated to Prof. Dr. Lev G. Magazanik, Sechenov Institute of Evolutionary Physiology and Biochemistry, RAS, St. Petersburg, Russia, at the occasion of his 80th anniversary.

ADVANCES IN INSECT PHYSIOLOGY VOL. 40
ISBN 978-0-12-387668-3
DOI: 10.1016/B978-0-12-387668-3.00001-5

Linyphiidae) and strongly neglected several other very abundant families (Araneidae, Lycosidae, Theridiidae, Thomisidae and Gnaphosidae).

We grouped the known 1618 records for venom compounds into six categories: low molecular mass compounds (16 % of all compounds), acylpolyamines (11 %), linear peptides (6 %), cysteine-knotted mini-proteins (60 %), neurotoxic proteins (1 %) and enzymes (6 %). Low molecular mass compounds are known from many spider families and contain organic acids, nucleosides, nucleotides, amino acids, amines, polyamines, and some further substances, many of them acting as neurotransmitters. Acylpolyamines contain amino acids (Araneidae and Nephilidae) or not (several other families) and show a very high diversity within one species. Linear peptides, also called cytolytic, membranolytic or antimicrobial, exert a highly specific structure and are so far only known from Ctenidae, Lycosidae, Oxyopidae and Zodariidae. Cysteine-knotted mini-proteins represent the majority of venom compounds because research so far focused on them. They probably occur in most but not all spider families. Neurotoxic proteins so far are only known from theridiid spiders. Enzymes had been neglected for some time but meanwhile it becomes obvious that they play an important role in spider venoms. Sixteen enzymes either cleave polymers in the extracellular matrix or target phospholipids and related compounds in membranes.

The overall structure of these compounds is given and the function, as far as it is known, is described. Since several of these component groups are presented in one average spider venom, we discuss the known interactions and synergisms and give reasons for such a functional redundancy. We also discuss main evolutionary pathways for spider venom compounds such as high variability among components of one group, synergistic interactions between cysteine-knotted mini-proteins and other components (low molecular mass compounds and linear peptides), change of function from ion-channel acting mini-proteins to cytolytic effects and replacement of mini-proteins by linear peptides, acylpolyamines, large proteins or enzymes. We also add first phylogenetic considerations.

1 Introduction

Spiders are after insects the most diverse terrestrial organism group and they comprise currently 42,055 species made available as an Internet database by Platnick (2011). Spiders evolved over some 300 million years, exert several characteristics and their unique combination is the reason for the unrivalled evolutionary success story of this group. (1) As hydraulic arachnids, spiders only possess flexor muscles in their limbs. Instead of extensor muscles, the hydraulic pressure of their hemolymph is fine-tuned by a well-balanced system of valves, which provides the necessary back-pressure. Since the long tubes of spider legs need only one system of muscles, spiders use the available space to enlarge their flexors, thus their muscles are relatively thicker than in comparable arthropods making spiders in general more powerful than insects (Shultz, 1991). (2) The distal end of the male pedipalp developed into a complex structure composed of fixed and movable sclerites that are used to transfer sperm to the female seminal receptacles during mating. These sclerites enable the males to perform safe sperm transfer, but also to lock this secondary sexual organ to the female genital opening, thus preventing mating outside the own species. This safe mode (key-lock-mechanism) is considered as major driver for a fast species radiation (Foelix, 2011). (3) Spiders possess up to six different silk gland types that allow them to use silk for a variety of needs. Spider silk is a polymer protein fiber with remarkable material properties, making it one of the most resistant and elastic biomaterials. Spiders developed various web types to capture prey items, they also wrap them with silk until they are defenceless, and they build silken retreats, cocoons and draglines. This makes spiders superior predators and there is presently only one spider species known feeding predominantly on plant material (Meehan *et al.*, 2009). (4) Spiders possess one pair of venom glands in the basal segment of the chelicerae (mygalomorph spiders) or in their prosoma (araneomorph spiders), and they inject small venom quantities into a prey item or predator to paralyze and/or kill and/or harm and/or possibly also to pre-digest it. Spider venom is a complex mixture of hundreds of components, mainly but not exclusively consisting of peptides and proteins, acting on myriad different targets in the prey organism. These targets include among others the extracellular matrix, membranes and a variety of receptors, quite often located in the nervous or muscular system.

The venom glands of spiders are already present in the oldest spider group, the Mesothelae, but they are very small. The cylindrical gland lies in the anterior portion of the cheliceral basal segment and a thin duct leads to an opening at the inner side of the fang, relatively distant from the tip of the fang (Foelix and Erb, 2010). Considering size and cheliceral power it may be doubtful how effectively the venom supports the predation success of these spiders. In mygalomorph spiders, the situation becomes clear. The well-developed venom glands, still in the basal cheliceral segment, produce powerful venom that is injected via the cheliceral fangs into the prey. The openings of the venom ducts lie at the convex side of the fang and can easily be seen if viewed from the ventral side. In all other spiders (Araneomorphae), the openings of the venom ducts are rather large and situated on the concave side of the cheliceral fangs, close to the tip. Moreover, the venom glands become much larger and reach into the prosoma where they can take a considerable proportion of this body part (Fig. 1; Foelix, 2011).

Though the possession of venom and venom glands is one of the key features of the success story of spiders, two or three spider families secondarily reduced their venom glands. (1) Uloboridae are small to medium-sized spiders producing cribellate silk. They lost their venom glands because uloborids evolved a unique technique to compact the living prey with extensive wrapping and by wetting the entire prey package with a powerful fluid that kills insects very fast. This fluid contains aggressive digestive enzymes dissolving the insect's intersegmental membranes and inner tissue, but it could also contain toxins (Weng *et al.*, 2006). (2) The only two species in the single genus *Holarchaea* of the family Holarchaeiidae, minute spiders from the moss and litter layer of the New Zealand wet

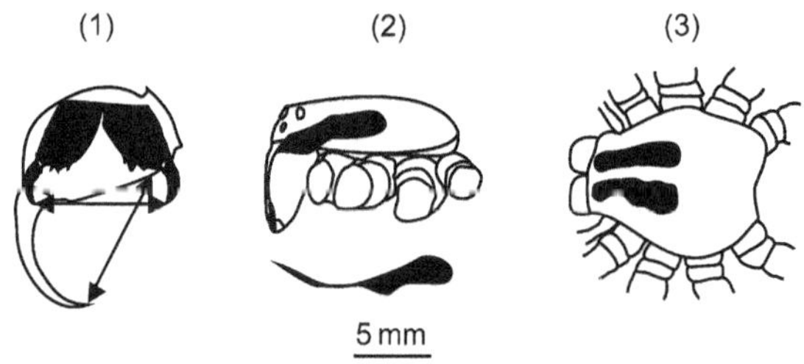

FIG. 1 (1) The functional superiority of a labidognath spider with small chelicerae (black) to an orthognath spider with large chelicerae (white) in subduing a prey item of the same size (arrow) is usually explained with such a scheme. Redrawn after Kaestner (1969). The position of the venom gland (black) in a labidognath spider (*Cupiennius salei*, Ctenidae), lateral (2) and dorsal view (3) (Malli *et al.*, 1993).

rain forest, have very long and slender cheliceral fangs without a venom gland opening on the cheliceral tip, and Forster and Platnick (1977) concluded that these highly modified spiders reduced their venom glands. (3) Symphytognathidae is another small family of tiny spiders, comprising about 65 species, characterized by chelicerae fused at least at their base. This raises doubts about the functionality of the chelicerae and it had widely been assumed that symphytognathids lost their venom glands. However, in a taxonomic review Forster and Platnick (1977) described that the fangs bear an opening on their retrolateral side that presumably leads to a poison gland. No venom gland loss but a complete change of function happened in spitting spiders (*Scytodes*, Scytodidae family, 228 species) that highly modified their venom glands to produce a gluey fibrous glycoprotein. This substance is spit from a distance of 2 cm well-directed onto the victim to fix it to the ground (Kovoor, 1987).

Even if the prey subduing process is much easier and safer for a spider when it can use its powerful venom, spiders do not exclusively rely on the use of their venom. The chelicerae of spiders are a strong tool to overwhelm a prey animal, and spiders routinely size and chew small arthropods without applying any venom. It is basically a question of the size ratio between predator and prey that determines the predator's decision: very small and/or defenceless arthropods are easily picked up and crashed with the chelicerae, while larger, dangerous or well-defended items are much more carefully approached and only attacked with venom injection. If the spider comes to the decision that the target is too large or dangerous, it will not attack and even get away.

Within the evolution of spiders the manageable size ratio of predators to prey has been considerably modified during several parallel processes. Most mygalomorph spiders are large animals; all have an orthognath chelicerae position and a relatively small venom gland compared to the body. The venom glands are restricted in size by the basal cheliceral segment, where muscles and venom glands compete for space. This limits the possibility to prey on large animals and mygalomorphs could only try to escape from this dilemma by becoming bigger, increasing the toxicity of their venom, or by constructing silken webs. The first two possibilities have been used up to some success but more or less functional webs can only be built by smaller species and are common only in a few mygalomorph groups (e.g. diplurids). The next most important evolutionary step was the change to a labidognath cheliceral position achieved by araneomorph spiders. This step was so successful that today more than 93% of all spider species are labidognath (Platnick, 2011). The modified cheliceral position does not need the big muscle packages of orthognath spiders; thus chelicerae could become smaller allowing parallel reduction of the body size. At the same time, the venom gland moved into the prosoma where it could enlarge its volume without conflicting too much with other organ systems (Fig. 1). These combinations allowed spiders to reduce their body size considerably, and today most spiders are small creatures with the most frequent body length size classes between 1 and 10 mm.

Within araneomorph spiders, sometimes also called ''true'' spiders, several lineages of webs evolved. Important achievements were the enlargement of webs into the third dimension (funnel webs, sheet webs, orb webs and many more types), regular and even geometrical arrangements of silken threats to reduce the amount of silk needed (up to orb webs that need only three attachment points and several other types of two-dimensional webs), the invention of cribellate silk or sticky glue to adhere entangled insects, thus preventing escape and reducing the need of a spider's immediate intervention (automated traps) and social mode of living with group-attacks on large prey items (Nentwig, 1985). These achievements, often combined, enable araneomorph spiders to predate on relatively large prey targets of all invertebrate groups (Nentwig and Wissel, 1986). Nevertheless, also labidognath spiders that do not build webs are very successful, and today some 50–55% of all labidognath spiders do not build webs while 45–50% are web-builders.

The natural prey of spiders consists predominantly of invertebrates, mainly insects. Among other arthropods, conspecific and non-conspecific spiders are a frequent part of a spider's diet. Further arthropod prey groups comprise opilionids and mites, uncommonly also millipedes and crustaceans (isopods). Among the non-arthropod invertebrates, occasionally annelid worms and slugs are eaten by spiders (Nentwig, 1987). Large spiders are also able to catch small vertebrates: Pisaurids are frequently successful in predating on small amphibians and small fish; mygalomorphs get occasionally small rodents and also birds, primarily small and helpless specimens such as nestlings; sparassids have casually been observed preying on small lizards.

Spiders are usually polyphagous, this means they are not specialised on a given prey type or taxon. Nevertheless, several cases of nutritional specialization among spiders are known. Bola spiders (*Mastophora* or *Celaenia* among Araneidae) lure their prey with the sex pheromone of noctuid moths and catch exclusively these moths (Gemeno *et al.*, 2000). Some mimetid genera (*Ero*, *Mimetus* and probably several more) are specialised web spider predators and developed complex behavioural tasks to overwhelm the resident web spider. *Zodarion* species and aphantochilid spiders are usually regarded as obligate myrmecophagous (Pekar *et al.*, 2007) and also in Theridiidae, Salticidae, Clubionidae and Gnaphosidae some genera specialized on ants. Spiders of the South African genus *Ammoxenus* (Ammoxenidae) are described as termite feeders (Dippenaar and Meyer, 1980). These records amount to roughly 2% of all known spider species that are specialized (i.e. more or less monophagous), while 98% are not.

From these considerations it can be concluded that the venom of spiders predominantly is not specific to a given invertebrate or insect group but rather unspecialized to be effective towards the broad spectrum of prey spiders naturally encountered. It is possible that the venom of the above listed monophagous spiders is more tailored towards their specific prey taxon but from none of the mentioned taxa venom has been investigated to date. Following the same line of argumentation, also vertebrates, including humans, are no important target of spiders and their venom. Nevertheless spider toxins are active against mammals including humans. One may hypothesize that the same mode of action is also effective or the same receptors/targets are also present in invertebrates and vertebrates or that the venom is complex enough to target a broad range of living organisms that may also be predators.

The effect of spider bites and spider venom on humans has largely been overestimated. Most spiders have only tiny mouthparts and short fangs, thus only large spiders can penetrate the strong human skin. Moreover, a specific aggressive behaviour or adaptation to the human environment is needed and there is consensus that only a limited number of taxa can seriously bite and harm humans. These spiders include several mygalomorph taxa (e.g. *Atrax* and *Hadronyche*) and the araneomorph genera *Phoneutria*, *Loxosceles* and *Latrodectus*, comprising altogether less than 100 species. The medical differential diagnosis of spider bites is usually performed not very professionally and includes other arthropod bites, skin infections, and exposure to chemical or physical agents. In total, there are about 100 reliably reported deaths from spider bites in the twentieth century, leaving spiders as mortality source of humans far behind bees and wasps or scorpions (Diaz, 2004; Isbister and Gray, 2002).

One of the strongest motivations to investigate spider venoms derives from this human argumentation. On the one hand spider venoms are of limited danger to humans, which justifies research into therapeutic means. On the other hand, spider venoms are seen as rich source of often highly specialized toxins, addressing selected receptors at muscles and nerve cells, also in vertebrates. This ''library'' or ''peptide treasure'' or ''natural goldmine'' of bioactives is being intensively investigated for scientific, medical and pharmaceutical reasons because some spider toxins proved to block highly specifically receptors playing a key role in important human physiological processes. There are recent reviews with a focus on pharmacological aspects of spider venoms (Corzo and Escoubas, 2003; Escoubas and Rash, 2004; Escoubas *et al.*, 2000b; Rash and Hodgson, 2002), on the importance of venom components as leads for new drugs (Bosmans *et al.*, 2009; Escoubas and King, 2009; Estrada *et al.*, 2007; Favreau *et al.*, 2001; Stöcklin and Favreau, 2002; Stöcklin and Vorherr, 2010), and, more generally, on biosynthesis, structure and function of venom compounds (Vassilevski *et al.*, 2009). Understanding the venomous function, how genes regulate such a complex machinery from chromosomes to venom gland to venom and what mechanisms are involved in the fast evolutionary process of venom components is also a major field of investigation (Ménez *et al.*, 2006).

With this review we provide a comprehensive and recent overview on the current diversity of venom compounds and relate it to the diversity of spiders. We have voluntarily decided to structure this review around structural patterns of venom components, using the mini-protein toxins (the largest family of spider toxins made of peptides strongly folded by disulphide bonds) as central element. We benefit from the progress in venom research of the recent years that included further spider species, thus enlarging our still limited knowledge basis a bit. We aim at providing a holistic view of the spider venom machinery as a concept, which not only enabled a spider to paralyze very efficiently most (if not all) potential prey items over time and adapt to predators and environmental changes, but which had also been modified and restructured several times within the evolution of these fascinating animals.

2 Methods and datasets

We intended to collect as comprehensively as possible information on compounds from spider venoms. Especially for mini-proteins, we extracted sequences from UniProt/SwissProt and GenBank, as well as from the specialized ''*Venoms*'' (Stöcklin and Cretton, 1998) and ''*ArachnoServer*'' (Herzig *et al.*, 2011) databases that contain information on the sequence, three-dimensional structure and biological activity of peptide and protein toxins derived from spider venoms. Dedicated bioinformatic tools from our integrated bioinformatics platform (Stöcklin and Cretton, 1998) and using recently developed software exploiting hidden Markov models and position-specific scoring matrix (PSSM, also known as generalized profiles) were used to extract, parse and classify the data. Over 200 entries describing fragments, partial sequences or that were not referenced have been discarded and our dataset has been processed to analyze mature sequences only (signal sequences were removed; latest access date 31.01.2011).

In addition we performed an intensive literature survey, starting with recent reviews such as Borges *et al.* (2009), Chesnov *et al.* (2001), Estrada *et al.* (2007), Foradori *et al.* (2005), Graudins *et al.* (2002b), Isbister and Gray (2003), Itagaki *et al.* (1997), Itagaki and Nakajima (2000), Kuhn-Nentwig (2003, 2009), McCormick and Meinwald (1993), Palma *et al.* (1997), Rash and Hodgson (2002), Liang (2004), Nagaraju *et al.* (2006), Rohou *et al.* (2007), Schroeder *et al.* (2008) and Vassilevski *et al.* (2009). Then we extended our search to further

TABLE 1
Number of compounds known from spider venom reviewed in this study

Substance group	Number of records	Number of species	Number of families
Low molecular mass compounds	259	85	26
Acylpolyamines	176	20	8
Linear peptides	102	17	8
Mini-proteins	965	60	20
Large proteins	11	7	1
Enzymes	105	49	14
Total	1618	174	32

primary literature and included finally 344 references. In total we aggregated 1618 data sets on venom compounds of spiders (Table 1). Since our literature records comprise more than 50 years in which spider taxonomy made considerable advances, we checked all spider names with Platnick's catalogue (Platnick, 2011), the authoritative taxonomic on-line catalogue for spiders. We consequently use only currently valid names of species but add a compilation of those species that appeared in the venom literature under older (invalid) names (Table 2).

3 Results: spider venom composition and modes of action

3.1 DISTRIBUTION OF RECORDS AMONG SPIDER GROUPS

Our data compilation yielded 174 spider species with at least one known venom component. Given the fact that as of January 31th 2011 42,055 spider species are known (Platnick, 2011), this amounts to 0.41% of all known species. These 174 spiders belong to 32 families corresponding to 29% of all existing spider families. Spiders investigated for venom research are not randomly distributed within the spider system but rather reflect a curious concentration on many big species (Hexathelidae and Theraphosidae among mygalomorph species, Nephilidae, Ctenidae, Sparassidae, some Lycosidae and Agelenidae) and the few species that are medically important for humans (e.g. the mentioned mygalomorphs and ctenids, as well as sicariids and theridiids; Fig. 2). The families with the highest species numbers (Salticidae and Linyphiidae) have been more or less ignored because these species are small and medically unimportant. In other very abundant families (Araneidae, Lycosidae, Theridiidae, Thomisidae, Gnaphosidae), each comprising more than 2000 species, research leading to structural elucidation of bioactive venom components focused so far only on very few species (Table 3).

TABLE 2
Taxonomic changes of spider names according to (1) progress in systematic research, (2) synonyms (usually an error during description as a new species), (3) wrong identification

Family	Invalid name	Valid name	Reason
Amaurobiidae	*Paracoelotes birulai*	*Pirenaitega luctuosa*	(1, 2)
Eresidae	*Eresus cinnaberinus*	*Eresus kollari*	(2)
	Eresus niger	*Eresus kollari*	(2)
Hexathelidae	*Atrax versutus*	*Hadronyche versuta*	(1)
Lycosidae	*Geolycosa raptoria*	*Lycosa erythrognatha*	(1, 2)
	Lycosa helluo	*Hogna helluo*	(1)
	Lycosa raptoria	*Lycosa erythrognatha*	(1, 2)
	Lycosa carolinensis	*Hogna carolinensis*	(1)
	Scaptocosa raptoria	*Lycosa erythrognatha*	(1, 2)
Oxyopidae	*Oxyopes kitabensis*	*Oxyopes takobius*	(3)
Sparassidae	*Isopeda immanis*	*Holconia immanis*	(1)
Theraphosidae	*Aphonopelma emiliae*	*Brachypelma emiliae*	(1)
	Brachypelma ruhnaui	*Brachypelma albiceps*	(2)
	Chilobrachys jingzhao	*Chilobrachys guangxiensis*	(2)
	Dugesiella hentzi	*Aphonopelma hentzi*	(1)
	Eurypelma californicum	*Aphonopelma hentzi*	(1, 2, 3)
	Grammostola cala	*Grammostola rosea*	(2)
	Grammostola spatulata	*Grammostola rosea*	(2)
	Haplopelma huwenum	*Haplopelma schmidti*	(1, 2)
	Pamphobeteus roseus	*Vitalius roseus*	(1)
	Ornithoctonus hainana	*Haplopelma hainanum*	(1)
	Ornithoctonus huwenum	*Haplopelma schmidti*	(1, 2)
	Plesiophrictus guangxiensis	*Chilobrachys guangxiensis*	(1)
	Pterinopelma vellutinum	*Vitalius vellutinus*	(1)
	Selenocosmia hainana	*Haplopelma hainanum*	(1)
	Selenocosmia huwenum	*Haplopelma schmidti*	(1, 2)
	Scodra griseipes	*Stromatopelma calceatum griseipes*	(1, 2)

Only those invalid names which appeared in the most recent venom literature are mentioned. For further material see Platnick (2011).

In depth venom research also concentrated only on a few species and the best-investigated spiders are the theraphosid *Haplopelma hainanum* (171 venom compounds known), the lycosid *Lycosa singoriensis* (154), the agelenid *Agelenopsis aperta* (86), two theraphosids *Chilobrachys guangxiensis* (80) and *Haplopelma schmidti* (68) and the two ctenids *Cupiennius salei* (65) and *Phoneutria nigriventer* (63). In nine other species more than 25 components are known. This concerns among mygalomorphs the hexathelids *Atrax robustus*, *Hadronyche versuta* and *Macrothele gigas* and the theraphosid *Stromatopelma calceatum griseipes*.

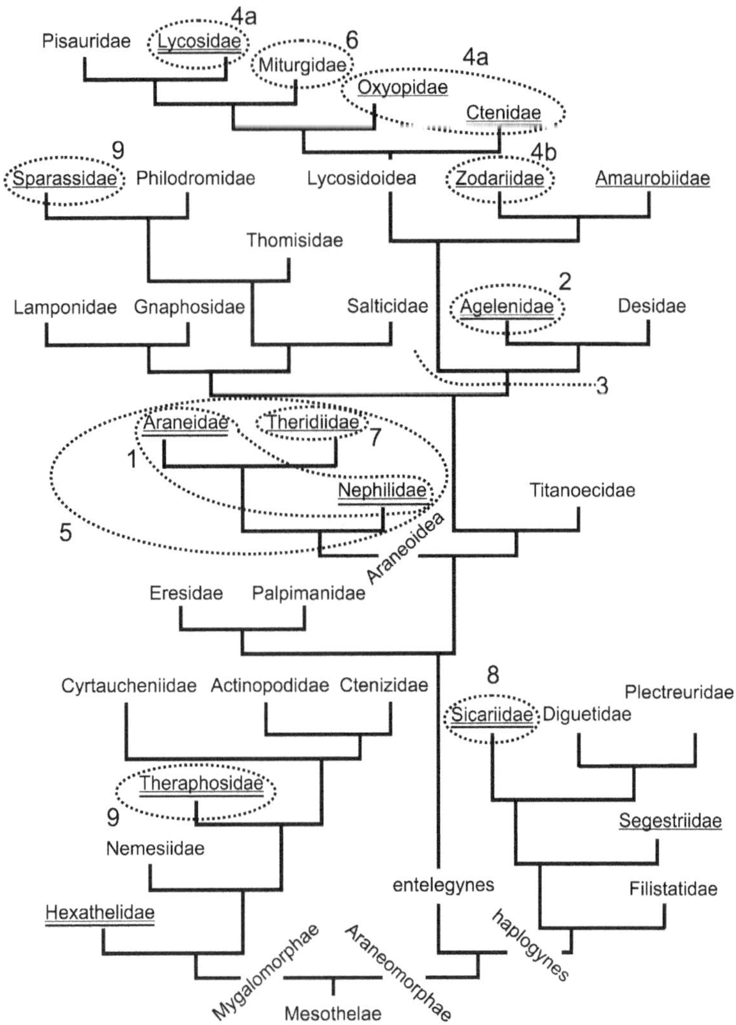

FIG. 2 Phylogenetic tree of the 32 spider families (out of a total of 109 families) from which at least one publication on a structurally elucidated venom compound is available. Modified from Jocqué and Dippenaar-Schoeman (2006) and Kuntner *et al.* (2008). *Underlined* family names refer to groups with > 10 entries (i.e. compounds and/or species) in our databases, *doubly underlined* families refer to groups with > 50 entries. (1) Exclusive presence of amino acid acylpolyamines, (2) non-amino acid acylpolyamines very common, (3) appearance of linear peptides, (4) presence of short and long cationic linear peptides (4a α-SCPs, 4b α-SCPs and α-LCPs), (5) absence of mini-proteins/cysteine-knotted peptides, (6) changed function of modified mini-proteins/cysteine-knotted peptides, (7) large proteins, (8) phospholipases D (PLDs) and (9) various neurotransmitter substances very common.

TABLE 3

Spider species occurring worldwide per family (species worldwide, all families with > 500 species included) and species included into this study (species investigated), arranged per family, decreasing order according to number of species known worldwide

Family	Species worldwide	Species investigated	Family	Species worldwide	Species investigated
Salticidae	5337	4	Oxyopidae	430	3
Linyphiidae	4378	0	Nemesiidae	353	2
Araneidae	3006	15	Miturgidae	348	2
Lycosidae	2374	13	Pisauridae	333	4
Theridiidae	2310	8	Amaurobiidae	276	2
Thomisidae	2146	3	Lamponidae	192	1
Gnaphosidae	2111	1	Desidae	182	1
Agelenidae	1146	7	Cyrtaucheniidae	134	2
Pholcidae	1111	0	Palpimanidae	131	1
Sparassidae	1109	10	Ctenizidae	125	1
Corinnidae	962	0	Sicariidae	124	19
Tetragnathidae	951	0	Filistatidae	113	1
Zodariidae	942	1	Segestriidae	111	3
Theraphosidae	933	39	Eresidae	96	3
Oonopidae	684	0	Hexathelidae	86	10
Clubionidae	570	0	Nephilidae	58	6
Dictynidae	565	0	Titanoecidae	49	1
Philodromidae	536	1	Actinopodidae	40	1
Anyphaenidae	514	0	Plectreuridae	30	1
Ctenidae	475	7	Diguetidae	15	1

Data on species abundance worldwide according to Platnick (2011).

Among araneomorphs, this includes the agelenid *Agelena orientalis*, the amaurobiid *Pirenaitega luctuosa*, the nephilid *Nephilengys borbonica*, the sicariid *Loxosceles intermedia* (25) and the zodariid *Lachesana tarabaevi* (28).

On the other side, from the venom of 53 spider species only one compound is known, there are 40 species with two compounds and 31 species with three described compounds. When considering only those families, where 20 or more compounds are known, this list comprises only 12 names: Agelenidae, Amaurobiidae, Araneidae, Ctenidae, Hexathelidae, Lycosidae, Nephilidae, Sicariidae, Sparassidae, Theraphosidae, Theridiidae and Zodariidae. It becomes obvious that our knowledge on spider venom is restricted to a few species and families (Fig. 2).

We aggregated the known 1618 records for venom compounds into six groups: low molecular mass compounds, acylpolyamines, linear peptides, mini-proteins, neurotoxic proteins and enzymes (Table 1).

3.2 COMPOUNDS OF LOW MOLECULAR MASS

We compiled 259 data sets concerning low molecular mass compounds in the venom of spiders. This refers to 85 species from 26 families with an average of 3 low molecular mass compounds known for a given species. Most compounds have been recorded for the theridiid *Latrodectus tredecimguttatus* (14 compounds), the hexathelid *Atrax robustus* (15 compounds), the ctenid *Cupiennius salei* and the theraphosid *Stromatopelma calceatum griseipes* (10 and 8 compounds; Table 4). Since most scientists investigating spider venom focus on peptides and proteins, not looking at low molecular mass compounds, we assume that many more substances have been detected but not published, or that the applied analytical methods prevented the detection and/or structural elucidation of low molecular mass compounds. Our conclusion is that probably most spider venoms contain a variety of low molecular mass compounds. There are only a few negative results: Schroeder *et al.* (2008) analysed the venom of three spiders (the diplurid *Phyxioschema* sp., the dysderid *Dysdera* sp. and the scytodid *Scytodes fusca*) with NMR spectroscopy and found only ''broad unresolved peaks characteristics of large peptides and proteins'', thus the venom was ''virtually devoid of any significant signals representing small molecules''. In the case of *Scytodes* this negative result is plausible because this genus highly modified its venom glands to a unique production of a gluey fibrous glycoprotein spitted at a distance of 2 cm on the victim (Kovoor, 1987). In the case of the other two mentioned species, the negative result may nevertheless be correlated to the applied technique.

3.2.1 *Ions, organic acids, nucleotides and nucleosides*

Spider venom has a specific ion configuration and also comprises a variety of small organic compounds. In principle most small molecules involved in metabolic and energy pathways of organisms, especially those concerning the venom production process in the venom glands, will also be present in spider venoms, even if only a few have been detected so far. Most low molecular mass compounds from venom can be seen in some context to neurotoxic substances or to neurons by supporting or enhancing other activities, that is they primarily support the toxicity of spider venom in a synergistic manner.

The venom cation concentrations of *Cupiennius salei* are 9 mM Na^+, 215 mM K^+ and 1 mM Ca^{2+}. This is in contrast to the cation concentrations in the hemolymph of this species (about 200 mM Na^+, 10 mM K^+ and 4 mM Ca^{2+}), so that the potassium concentration in the venom is increased more than 20-fold, compared to the hemolymph, and the sodium concentration is decreased, respectively (Kuhn-Nentwig *et al.*, 1994). A comparable ion composition is known from a tarantula venom with 9.4 mM Na^+, 70.4 mM K^+, 2.9 mM Ca^{2+} and 11.2 mM Mg^{2+} (Savel-Niemann and Roth, 1989). At such concentrations, potassium is able to induce depolarization of excitable cell membranes causing paralysis of a prey item. Potassium is also known as effective synergist of neurotoxins in *Cupiennius salei*. (Wullschleger *et al.*, 2004, 2005). Such cation concentrations and their synergistic effect on mini-proteins are only known from *Cupiennius salei*, but we assume that this is a wider-spread situation among spiders and possibly many other venomous organisms. This assumption is supported by a comparable mechanism that has been discussed for the prevenom of the scorpion *Parabuthus transvaalicus* (Inceoglu *et al.*, 2003).

One fifth of all records of low molecular compounds (20%) concerns citric acid, detected in 48 species from 16 families (Table 4). There are at least four possible reasons for the presence of citric acid in spider venoms. (1) Citric acid alone is effective in preventing bacterial growth at concentrations (6.25–25 mg mL^{-1}; Lee *et al.*, 2001) occurring in different spider venoms (16–147 mM). Interestingly, citric acid is also present in the venom of snakes, bees and scorpions (Fenton *et al.*, 1995). (2) It could be assumed that citric acid serves as an enhancer to enforce the effect of antimicrobial or other substances (Shanbrom, 2003). It has not yet been shown that citric acid also enhances mini-proteins. We did not detect a simple co-occurrence of cytolytic (antimicrobial) substances and of citric acid in our database, since both compounds co-occur only in the venom of five spider families, whereas in 3 more families only cytolytic substances and in 11 more families only citric acid has been detected. Furthermore, most of such studies tend to focus on one family of compounds and it is thus difficult to conclude. (3) Citric acid as divalent metallic ion chelator may partially inhibit zinc ion metalloproteases and Ca^{2+}-dependent enzymes (i.e. phospholipases A_2) in the venom glands. After injection into a prey item, the high citric acid concentration could be diluted and the enzymes are activated. This has been postulated for bee and snake venoms (Odell *et al.*, 1998, 1999) and could also be important in spider venoms. Our proof of such a relationship, however, is not very strong: according to our database, in seven spider families citric acid and enzymes have been found to occur together, in nine families only citric acid and in seven families only

TABLE 4
Records of low molecular mass compounds according to spider family and species

Family	Genus	Species	Citric acid	Glutamic acid	Acetyl-choline	GABA	5-Hydroxytryptamine
Agelenidae	*Agelena*	*labyrinthica*					
	Agelena	*orientalis*					
	Agelenopsis	*aperta*	x				
	Hololena	*curta*	x				
	Tegenaria	*agrestis*	x				
	Tegenaria	*domestica*	x			x	
Amaurobiidae	*Coelotes*	*pastoralis*	x				
	Pirenaitega	*luctuosa*	x				
Araneidae	*Aculepeira*	sp.	x				
	Araneus	*diadematus*	x	x			
	Araneus	*gemma*		x			
	Araneus	*tartaricus*	x	x			
	Eriophora	*edax*	x	x			
	Larinioides	*cornutus*	x	x			
	Parawixia	*bistriata*					x
Ctenidae	*Cupiennius*	*salei*	x	x			
	Phoneutria	*fera*					x
	Phoneutria	*nigriventer*				x	
Cyrtaucheniidae	*Anemesia*	*incana*		x			
Eresidae	*Eresus*	sp.		x			
	Stegodyphus	sp.		x			
Filistatidae	*Kukulcania*	*hibernalis*				x	
Gnaphosidae	*Drassodes*	sp.	x	x			
Hexathelidae	*Atrax*	*robustus*	x			x	x
	Macrothele	sp.				x	
Lamponidae	*Lampona*	*cylindrata*					
Lycosidae	*Hogna*	sp.	x	x			
	Lycosa	*erythrognatha*					x
	Lycosa	*godeffroy*					
	Lycosa	*indagastrix*		x			
	Lycosa	*proegrandis*	x	x			
	Rabidosa	sp.	x	x			
Nemesiidae	*Raveniola*	*micrura*	x	x		x	

Noradrenaline	Histamine	Octopamine	Spermine	Spermidine	Cadavrine	Putrescine	ATP, ADP, AMP	Nucleosides	Others
								x	
								x	
								x	
x	x							x	
	x								
		x						x	
	x							x	
								x	
								x	
									Choline
									Adrenaline, dopamine, epinephrine, epinine, 3,4-dihydroxy-phenylacetic acid, aspartic acid
								x	
								x	Adenosine, guanosine, inosine, 2-amino-5-ureido-pentanamide, 2-3-4-trihydroxypurine, 6-hydroxytryptargine, 1-4-guanidinobutoxy-6-hydroxy-1,2,3,4-tetrahydro-b-carboline, 1-guanidino-6-hydroxy-3,5-dihydro-b-carboline
	x				x	x	x		Aspartic acid, taurine
	x								
		x	x	x	x	x			Glycerol, glycine, lactic acid, phosphoric acid, tyramine, urea
								x	
x	x								
								x	
	x								
	x								
	x								
	x								
		x						x	

(*continues*)

TABLE 4 (*Continued*)

Family	Genus	Species	Citric acid	Glutamic acid	Acetyl-choline	GABA	5-Hydroxytryptamine
Oxyopidae	*Peucetia*	*viridans*		x			
Palpimanidae	*Palpimanus*	sp.	x				
Philodromidae	*Thanatus*	*striatus*		x			
Pisauridae	*Dolomedes*	*gertschi*	x	x		x	
	Dolomedes	*tenebrosus*	x	x			
	Pisaura	*mirabilis*		x		x	
Plectreuridae	*Plectreuris*	*tristis*		x			
Salticidae	*Phiddipus*	sp.	x				
Segestriidae	*Ariadna*	sp.	x				
Sicariidae	*Loxosceles*	*arizonica*					
	Loxosceles	*deserta*					
	Loxosceles	*intermedia*					
	Loxosceles	*reclusa*	x				
Sparassidae	*Delena*	*cancerides*	x	x			
	Eusparassus	*oculatus*		x	x	x	
	Holconia	*flindersi*	x	x		x	
	Isopedella	*canberrensus*	x	x			
	Isopedella	*leai*	x			x	
	Micrommata	*virescens*	x	x	x		
	Neosparassus	*diana*	x	x		x	
	Olios	sp.		x			
Theraphosidae	*Acanthoscuria*	*atrox*					x
	Acanthoscuria	*sternalis*					x
	Aphonopelma	*hentzi*	x	x		x	
	Aphonopelma	*seemani*	x		x	x	
	Aphonopelma	sp.	x			x	
	Avicularia	*avicularis*		x		x	
	Brachypelma	*emiliae*	x				
	Brachypelma	*smithi*					
	Ceratogynus	*marshalli*	x			x	
	Chilobrachys	sp.	x		x		
	Citharischius	*crawshayi*	x	x			
	Grammostola	*rosea*	x				
	Haplopelma	*albostriatum*	x	x	x		
	Haplopelma	*hainanum*	x	x	x	x	
	Haplopelma	*lividum*	x	x	x		
	Haplopelma	*schmidti*	x	x	x	x	
	Heteroscodra	*maculata*				x	
	Lasiodora	*klugii*					x
	Phormictopus	*cancerides*	x			x	
	Psalmopoeus	*cambridgei*	x			x	
	Stromatopelma	*calceatum griseipes*	x			x	
	Vitalinus	*vellutinus*					x
	Vitalius	*roseus*					x
Theridiidae	*Latrodectus*	*menavodi*					
	Latrodectus	*tadzhicus*					
	Latrodectus	*tredecimguttatus*		x		x	x
	Steatoda	*paykulliana*		x		x	
Thomisidae	*Misumena*	*vatia*	x			x	
	Thomisus	*onustus*				x	
Titanoecidae	*Titanoeca*	sp.	x	x			
Zodariidae	*Lachesana*	sp.	x	x		x	

Noradrenaline	Histamine	Octopamine	Spermine	Spermidine	Cadavrine	Putrescine	ATP, ADP, AMP	Nucleosides	Others
								x	
	x								
									Choline
								x	
								x	
	x								
								x	
		x						x	
	x	x						x	
			x	x	x	x	x		Aspartic acid
								x	
			x	x	x	x	x	x	
								x	
			x	x	x	x			
		x							
								x	
								x	
								x	
	x		x					x	Adenosine
		x							
			x	x	x	x			Lactic acid, phosphoric acid,
									Adenosine, guanine, inosine, 2,4,6-trihydroxypurine
								x	Aspartic acid
								x	
								x	
								x	

enzymes have been found. (4) Citrate could serve as negative counter ion for the highly cationic cytolytic peptides and acylpolyamines in the venom of spiders and may be responsible for pH values between 5.3 and 6.1 (Kuhn-Nentwig *et al.*, 1994; Savel-Niemann and Roth, 1989; Schanbacher *et al.*, 1973a). Taking again the fact that most spider venoms are not yet carefully investigated for most of the here mentioned compounds, no final conclusion can yet be drawn onto this topic. Other organic acids have been detected only occasionally in spider venoms. Records include lactic acid and phosphoric acid from theraphosid and hexathelid spiders (Duffield *et al.*, 1979; Lange *et al.*, 1992).

Sulphated (six records) and other nucleosides (25 records) have been detected in the venom of 30 species from 11 families (Schroeder *et al.*, 2008). The first fucosylated guanosine disulphate HF-7 (Fig. 3) was identified in the venom of *Hololena curta* that effectively blocks kainate receptors and to a less extent L-type calcium channels (McCormick *et al.*, 1999). In the venom of *Loxosceles arizonica*, *L. deserta*, and *L. reclusa* further sulphated nucleosides (2,5-disulfated guanosine as major component) even represent by far the most abundant components. A comparable situation has been reported for the agelenid *Tegenaria agrestis* where sulphated nucleosides constitute about 50% of its venom total dry mass (Taggi *et al.*, 2004). Nucleosides like adenosine, guanosine, 2,3,4-trihydroxypurine and inosine are identified in different combinations in several spider families (Horni *et al.*, 2001; Rodrigues *et al.*, 2004; Schroeder *et al.*, 2008). In this context, the venom of the colonial spider *Parawixia bistriata* is spotlighted to its high content of different low molecular mass compounds. Rodrigues *et al.* (2004) identified inosine in this venom and could demonstrated that the analogue inosine 5′-monophosphate induced a delayed paralytic effect in termite soldiers and workers. Remarkable is the further characterisation of tetrahydro-β-carboline toxins from *P. bistriata*: PwTX-I (Fig. 3; 6-hydroxytrypargine, LD_{50} 8 ng g^{-1} honeybee; Cesar *et al.*, 2005b) and PwTx-II (1-4-guanidinobutoxy-6-hydroxy-1,2,3,4-tetrahydro-β-carboline, LD_{50} 12 ng mg^{-1} honeybee; Cesar *et al.*, 2005a) acting in low doses insecticidal and to rats neurotoxic, convulsive and lethal. Saidemberg *et al.* (2009) could demonstrate that these indolalkaloid toxins are non-competitive inhibitors of monoamine oxidase A and B.

Nucleotides such as ATP, ADP and AMP are so far known from the venom of three spider species belonging to Theraphosidae and Ctenidae. Chan *et al.* (1975) reported a high content of adenosine 5′-triphosphate identified in the venom of *Aphonopelma* sp. (56.6 μg μL^{-1} venom) and *Aphonopelma hentzi* (28.1 μg μL^{-1} venom) enhancing at a non-toxic level the activity of ''venom necrotoxin'' isolated from *A. hentzi* (Chan *et al.*, 1975). These substances are part of energy delivering pathways and can therefore support the effect of other compounds, especially mini-proteins (Chan *et al.*, 1975; Savel-Niemann and Roth, 1989). It is possible that they occur in all spider venoms but only rarely have been confirmed.

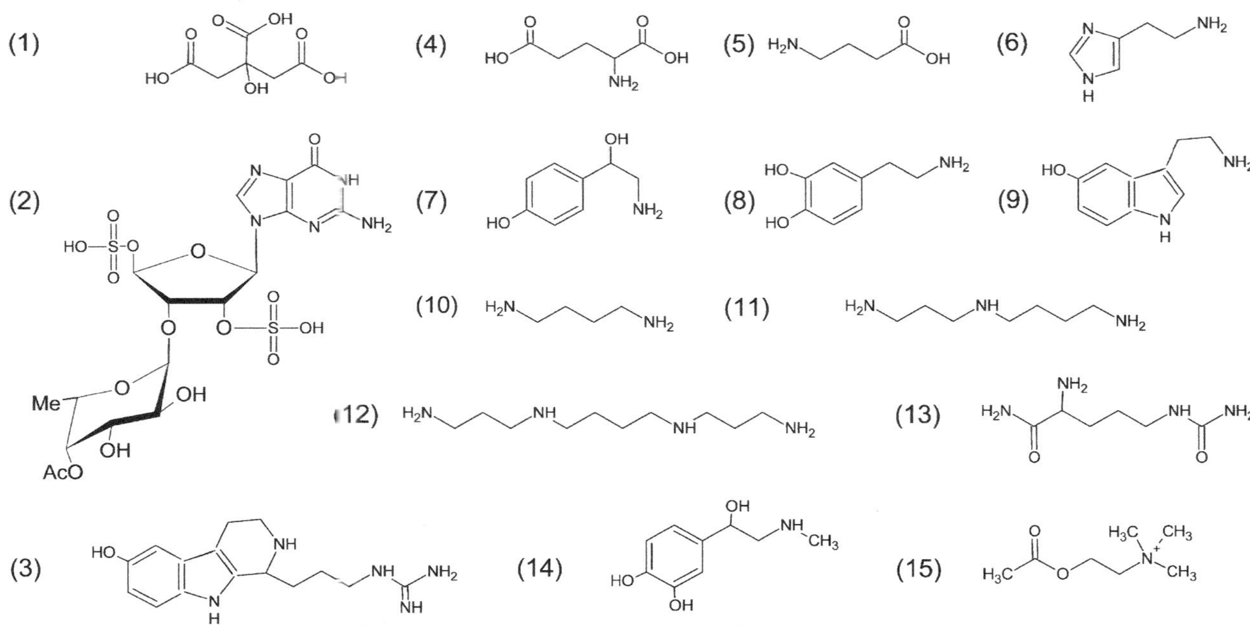

FIG. 3 Chemical structures of low molecular mass substances identified in the venom of spiders: (1) citric acid, (2) HF-7 (Taggi *et al.*, 2004), (3) PwTx-I (Cesar *et al.*, 2005b), (4) glutamic acid, (5) γ-aminobutyric acid, (6), histamine, (7) octopamine, (8) dopamine, (9) serotonin, (10) putrescine, (11) spermidine, (12) spermine, (13) FrPbAII (Gelfuso *et al.*, 2007), (14) adrenaline and (15) acetylcholine.

3.2.2 *Amino acids, amines, polyamines, acetylcholine and noradrenalin*

More than half of all records of low molecular mass compounds (56%) concern substances functioning as neurotransmitters and neuromodulators:

(1) amino acids;
(2) biogenic amines and polyamines and
(3) other neurotransmitters such as acetylcholine and noradrenalin.

These compounds have more or less equally been reported from 25 mygalomorph, haplogyne and entelegyne spider families. However, the venom of all eight Sparassidae and of 60% of all investigated 37 Theraphosidae contained neurotransmitters. This could indeed indicate that these two families rely more than the other groups on neurotransmitters.

3.2.2.1. Amino acids Most if not all biogenic amino acids are expected to occur regularly in spider venoms but have hardly been reported. In the venom of *Cupiennius salei*, Kuhn-Nentwig *et al.* (1994) found 20 biogenic amino acids in total. In addition, taurine has been reported from the venoms of *C. salei* where it occurs in high amounts and from the black widow *Latrodectus tredecimguttatus* (Bettini and Maroli, 1978; Kuhn-Nentwig *et al.*, 1994). Usually amino acid sequencing machines do not detect taurine since it is not included into the standard amino acid analyses procedure, and therefore, it can certainly be assumed that it is wider spread than currently known. Taurine is common in the nervous system of spiders and insects where it has an inhibitory function and is involved into the control of hyperexcitability of the central nervous system (Meyer, 1991; Strang *et al.*, 1990). Besides GABA (γ-aminobutyric acid), taurine acts as agonist of ligand-gated chloride channels through activation of $GABA_A$ receptors (Moore *et al.*, 2006). Taurine alone, however, has no own toxicity towards the blowfly *Protophormia* but increases at venom concentration the potency of CSTX-1 (ω-ctenitoxin-Cs1a), the main neurotoxin in the venom of *C. salei* (Kuhn-Nentwig *et al.*, 1998). Remarkably, coinjection of taurine and CSTX-1 into *Drosophila melanogaster* exhibits no increasing effect on the activity of the neurotoxin (Wullschleger *et al.*, 2005). Maybe this result is caused by different taurine sensitivities of insects as the concentrations in the hemolymph of the blowfly (464 pmol μL^{-1}) is threefold higher when compared with *D. melanogaster* (160 pmol μL^{-1}; Kuhn-Nentwig *et al.*, 1998). Nevertheless, the parasitoid wasp *Ampulex compressa* injects a cocktail of GABA, taurine and β-alanine directly into the prothoracic ganglion of the cockroach *Periplaneta americana*, which serves as food source for its offspring. This mixture induces in the cockroach a transient paralysis of the front legs by opening chloride channels for central synaptic block (Moore *et al.*, 2006).

The occurrence of glutamate (39 records) and GABA (28 records) in the venom of spiders points to its relevance to the envenomation process of a given prey together with other spider venom compounds. Both compounds are widely spread among the investigated spider species and recorded from 20 families. Glutamate is present in the venom from less than 0.02 nmol μL^{-1} (Kuhn-Nentwig *et al.*, 1994) up to 425 nmol μL^{-1} (Early and Michaelis, 1987) and acts in insects as an excitatory neurotransmitter at the neuromuscular junctions targeting glutamate receptors (GluRs). Activation by glutamate results in a cation permeability of these receptors. In contrast, in the central synapses of insects, glutamate and GABA act as inhibitors of glutamate and GABA receptors. After activation, both postsynaptic receptors become permeable to chloride ions (El Hassani *et al.*, 2008; Narahashi *et al.*, 2007).

3.2.2.2. Biogenic amines and polyamines Many biogenic amines, polyamines and their receptors exert important functions in modulation and regulation of a great variety of physiological and behavioural processes in both invertebrates and vertebrates. In invertebrates, five major biogenic amines have been identified: histamine, octopamine, tyramine, serotonin and dopamine (Blenau and Baumann, 2001), which were also detected in different spider venoms (Fig. 3).

Histamine (14 records) has been identified as neurotransmitter and neuromodulator in vertebrates (Parsons, 1991), as well as in invertebrates (Hardie, 1987). In arthropods, histamine acts as neurotransmitter of photoreceptors (Nässel, 1999) and mechanoreceptors (Buchner *et al.*, 1993) by directly activating cys-loop ligand-gated chloride channels (Hardie, 1989). With the application of neurophysiological, immuno-histochemical, molecular biological and bioinformatical methods histamine-gated chloride channels or subunits have been identified in many insects such as *Drosophila melanogaster* (Gisselmann *et al.*, 2002; Witte *et al.*, 2002; Yusein *et al.*, 2010), the grasshopper *Locusta migratoria* (Janssen *et al.*, 2010), the parasitoid wasp *Nasonia vitripennis* and the honey bee *Apis mellifera* (Jones *et al.*, 2010), and in many more dipteran, hymenopteran, coleopteran and lepidopteran species (Iovchev *et al.*, 2006).

The content of histamine in various spider venoms differs considerably: *Loxosceles intermedia* exhibits the highest concentration with 89 ng μg^{-1} crude venom (Paludo *et al.*, 2009), followed by *Lampona cylindrata* with 50 ng μg^{-1} crude venom (Rash *et al.*, 2000), *Lycosa godeffroyi* 44.5 ng μg^{-1} crude venom (Rash *et al.*, 1998) and *Lycosa erythrognatha* (Fischer and Bohn, 1957) with 36 ng μg^{-1} crude venom. In contrast, the ctenids *Phoneutria nigriventer* (6–10 ng μg^{-1} crude venom; Schenberg and Pereira Lima, 1978) and *Cupiennius salei* (4.5 ng μg^{-1} crude venom; Kuhn-Nentwig *et al.*, 1994) possess only one twentieth of the concentration of *Loxosceles intermedia*. Paludo *et al.* (2009) stated that the high amount of histamine within this venom is able to cause edema and an increase in vascular permeability when applied in mice.

Histamine is known to occur in the venom of 14 species from nine spider families and it is probable that it may be present in much more species. Besides the above-mentioned neurotransmitter function, the discussion on the meaning of histamine in spider venoms covers three further topics:

(1) It produces pain and therefore it is seen as a defensive substance. The argumentation for the "pain theory" comes mainly from hymenopteran venoms containing high amounts of histamine and that is used by bees and wasps to defend against vertebrate predators (Bettini, 1978). In the case of spiders this argumentation is not convincing because only large spiders (Ctenidae, Sparassidae, Theraphosidae) suffer from vertebrate predators while for most smaller spiders invertebrates are more important predators.
(2) Histamine also increases local blood flow and cell permeability which facilitates spreading of the venom in the tissue of the prey. Since prey of spiders consists almost exclusively of invertebrates, the "blood flow theory" is also a vertebrate targeted argumentation perhaps at most concerning a few mygalomorph spiders occasionally taking small vertebrates as prey.
(3) The fact that histamine acts in insects as neurotransmitter targeting ionotropic receptors channels pointed to a possible role as modulator after envenomation of insects. Kuhn-Nentwig *et al.* (1994) and Wullschleger *et al.* (2005) clearly demonstrated that histamine enhances in a synergistically manner the main mini-protein CSTX-1 (ω-ctenitoxin-Cs1a) from the venom of *Cupiennius salei* in a *Drosophila* bioassay. Interestingly, the insecticidal effect of the cytolytically acting peptide Cu 1a (M-ctenitoxin-Cs1a) and of the enhancer peptide CSTX-13 (U_2-ctenitoxin-Cs1a) was not supported by co-injection of histamine.

Octopamine (seven records) and tyramine (one record), two biogenic phenolamines are involved as independent neurotransmitters in multiple physiological functions of invertebrates, mainly as antagonistic modulators of behaviour and metabolism (Blenau and Baumann, 2001). Both are present in the nerve system of insects and are decarboxylation products of the amino acid tyrosine with tyramine being the biological precursor of octopamine. These two monoamines are the counterpart to the structurally related vertebrate transmitters/hormones adrenaline and noradrenaline, acting through G protein-coupled receptors (Roeder, 2005). Octopamine is identified in many arthropods, it is considered as a "fight or flight" hormone in insects, and it plays a role in the modulation of flight muscles (Roeder, 1999, 2005). Octopamine is known from the venom of seven spider species belonging to five families (Agelenidae, Hexathelidae, Nemesiidae, Sparassidae and Theraphosidae). The concentration of octopamine in the venom of *Atrax robustus* is determined with 3.7 ng μg^{-1} female venom

(Duffield *et al.*, 1979). Beside other effects, tyramine modulates in *Drosophila melanogaster* the locomotion of larvae (Saraswati *et al.*, 2004). Up to now, tyramine has only been identified in the venom of *Atrax robustus* (0.61 ng μg^{-1} female venom; Duffield *et al.*, 1979).

Dopamine (one record) and 5-hydroxytryptamine (5-HT; serotonin; 10 records) are well known as neurotransmitters and modulators in the invertebrate and vertebrate nervous systems (Scheiner *et al.*, 2006). Dopamine was only reported from the venom of *Araneus gemma* (4.3 nM in venom; Early and Michaelis, 1987), whereas 5-HT was identified in the venom of *Latrodectus tredecimguttatus* (Pansa *et al.*, 1972), six species of Theraposidae, *Atrax robustus*, *Lycosa erythrognatha* and *Phoneutria fera*, which exhibits with 2.5 ng μg^{-1} dried venom the highest concentration. Comparable to histamine, 5-HT is able to produce severe pain in vertebrates (Welsh and Batty, 1963).

Polyamines (Fig. 3) in spider venoms comprise spermine (six records), spermidine (five records), putrescine (six records) and cadaverine (six records) and they are so far known from seven spider species belonging to three families (Ctenidae, Hexathelidae, Theraphosidae; Cabbiness *et al.*, 1980; Duffield *et al.*, 1979; Kuhn-Nentwig *et al.*, 1994; Lange *et al.*, 1992; Moore *et al.*, 2009; Savel-Niemann, 1989). Generally, polyamines with the exception of cadaverine, are important in protein synthesis, cell division and cell growth, and they are present in nearly all cells unbound in a concentration range from 10 to 100 μM. Forsythe (1995) suggested that invertebrate venom polyamines may block different voltage- and ligand-gated ion channels and GluRs by a comparable mechanism to the endogenous counterparts, but with relative irreversible effects. Intracellular spermine blocks inward rectifier K^+ channels and AMPA/kainate receptors, which seems to control the resting membrane potential and excitability in many cell types. Extracellular spermine modulates and blocks NMDA receptors affecting the excitability and influx of Ca^{2+} ions in different cells (neurons, glial cell) of the nervous system (Williams, 1997a,b). Meanwhile, three cDNA encoding inward rectifier K^+ channels were isolated from *Drosophila melanogaster* (Döring *et al.*, 2002; MacLean *et al.*, 2002) and it is tempting to speculate a possible role of polyamines in blocking such kind of channels in insects, and in the case of mygalomorph spiders also in small vertebrates.

Additionally, CNG channels and HCN channels, both belonging to cyclic nucleotide-gated cation channels (Biel, 2009), are present in many mammalian tissue types and are blocked in different magnitudes by intracellular and extracellular spermine, spermidine and putrescine (Guo and Lu, 2000; Vemana *et al.*, 2008). A putative CNG channel and one gene (DMIH) coding for HCN channels have been identified in *Drosophila* (Ishii *et al.*, 2007; Miyazu *et al.*, 2000), which could point at possible targets for polyamines. Remarkable is a further polyamine, FrPbAII (2-amino-5-ureido-pentanamide) from the venom of *Parawixia bistriata*, which inhibits synaptosomal uptake of GABA and glycine and seems to be a potent anticonvulsant for rats (Beleboni *et al.*, 2006; Gelfuso *et al.*, 2007; Liberato *et al.*, 2006).

3.2.2.3. Acetylcholine, choline, adrenaline and noradrenaline In the insect nervous system, acetylcholine is the most abundant excitatory neurotransmitter of chemo- and mechanosensory receptor neurons. The nicotinic acetylcholine receptors (nAChRs) are most numerous and mediating fast synaptic transmission. Nevertheless, muscarinic acetylcholine receptors (mAChR) are also identified in insect brain (Homberg, 2002). Acetylcholine was identified in the venoms of sparassid (two records) and theraphosid spiders (five records) and choline (two records) only in one araneid and one segestriid species (Schroeder *et al.*, 2008). Noradrealine (two records from an agelenid and a lamponid spider) and adrenaline (one record from an araneid spider) seem less abundant in spider venom (Early and Michaelis, 1987; Frew *et al.*, 1994; Rash and Hodgson, 2002).

3.3 ACYLPOLYAMINES

Acylpolyamine are a class of neuroactive compounds, containing a polyamine structure that mainly contains an aromatic acyl endmoiety (indolic or phenolic) at its end, with molecular masses between 350 and 1000 Da. A high structural diversity within these compounds is obtained by a combination of different acyl groups with polyamine chains, varying in length, number of amide bonds and functional groups (Itagaki and Nakajima, 2000; Schäfer *et al.*, 1994). There are two groups of acylpolyamine toxins, those containing amino acids and those not containing them, but their effect on the GluR is similar (Fig. 4; Itagaki and Nakajima, 2000; McCormick and Meinwald, 1993).

Acylpolyamines are characterised by positive charges on the free and secondary amines of their polyamine chain, which explains their blocking activity on cationic-selective ion channels. In general, these toxins are active on Ca^{2+} channels, K^+ channels and ionotropic receptors (Mellor and Usherwood, 2004). Starting investigations in the 1980s with venom or isolated acylpolyamine toxins from *Nephila* (Kawai *et al.*, 1982; Toki *et al.*, 1988), *Argiope* (Budd *et al.*, 1988; Grishin *et al.*, 1989), *Agelenopsis* (Skinner *et al.*, 1989) and *Araneus* species (Early and Michaelis, 1987), such compounds were identified as specific blockers of glutamatergic synapses (Kawai *et al.*, 1982; Usherwood *et al.*, 1984). In insects, their biological activity is characterized by a reversible paralysis caused by use-dependent blocking of activated postsynaptic GluRs channels (Adams, 2004; Usherwood, 1994; Usherwood *et al.*, 1984). The binding site of these non-competitive inhibitors is located within the pore of the GluR and has to be accessible by ligand activation. Only in the presence of synaptic activity, polyacylamine toxins interact with the GluR (Usherwood, 1994; Usherwood and Blagbrough, 1989). Interestingly, orb-weaving spiders (Araneidae and Nephilidae) exhibiting acylpolyamine toxins in their venom also possess glutamate for receptor activation which explains the efficacy of this venom composition (Early and Michaelis, 1987; Schroeder *et al.*, 2008). The concentration of glutamate in milked venom of *Araneus gemma* varies between

(1)

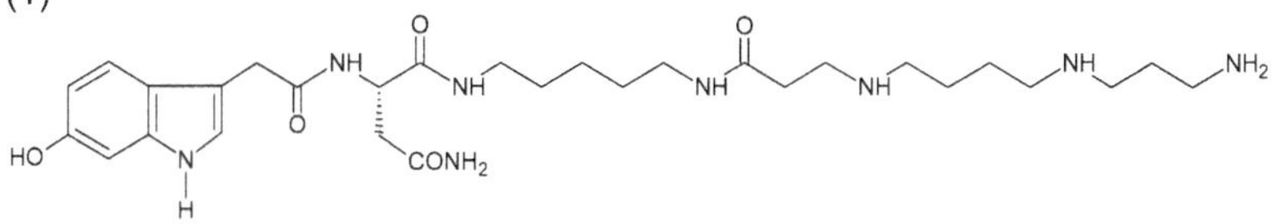

(2)

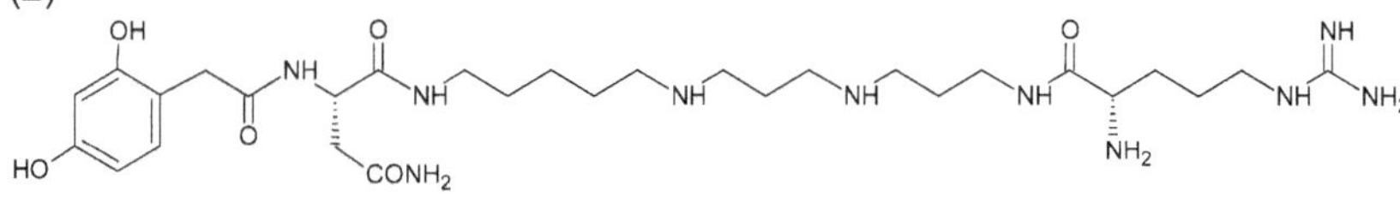

(3)

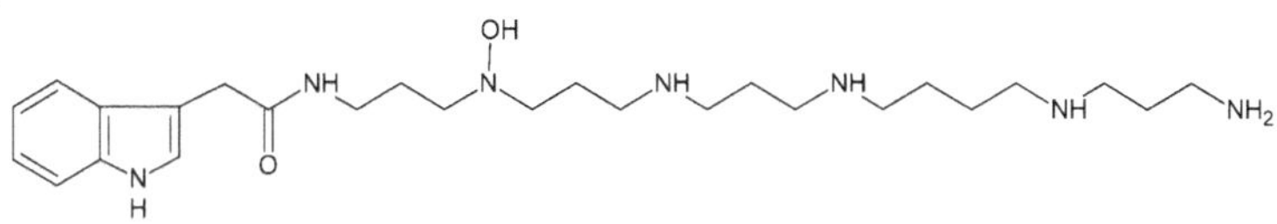

(4)

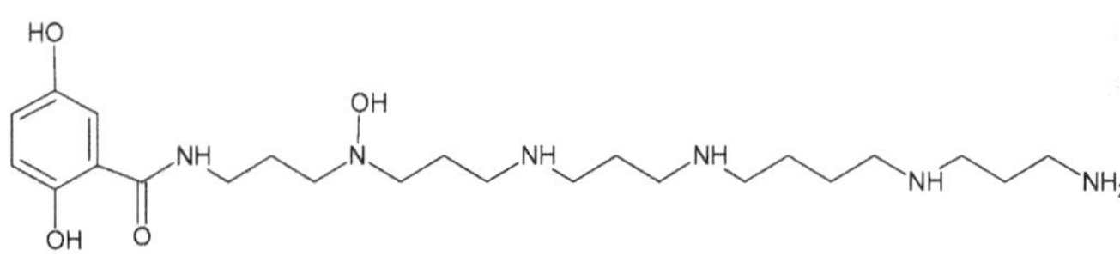

FIG. 4 Characteristic acylpolyamine toxins from spider venom. (1) Amino acid-containing acylpolyamines exhibiting an indolic NPTX-1 (Toki *et al.*, 1988) or phenolic end group (2) Argiopine (Grishin *et al.*, 1989), connected via asparagine to the polyamine chain; and (3) non-amino acid-containing acylpolyamines exhibiting an indolic AGEL 489 or phenolic end group (4) AGEL 468 (Jasys *et al.*, 1990) connected to the polyamine chain.

130 and 425 mM (Early and Michaelis, 1987). Another activation strategy is identified in the venom of Agelenidae, which possess no glutamate in the venom, but stimulate release of glutamate from pre-synaptic terminals by μ-agatoxins acting as voltage-activated sodium channel modifier (Adams *et al.*, 1989). The concentration of some acylpolyamine toxins (Fig. 4) has been estimated in native venom of *Agelenopsis aperta* and ranges from 1.36 (Agel 468) to 33.02 μg μL^{-1} venom (Agel 489; Jasys *et al.*, 1990).

Although spiders developed acylpolyamine toxins for rapid paralysis of their invertebrate preys, these toxins block acetylcholine receptors and GluRs in vertebrate nervous system in sub- and micro-molar concentrations. Beside other targets, capsaicin receptor channels, which play an important role in temperature sensing and nociception of vertebrates are inhibited by some of these acylpolyamines isolated from the venom of *Agelenopsis aperta* (Kitaguchi and Swartz, 2005).

Remarkably, in the venom of one species many different acylpolyamine toxins were detected and their structures could be elucidated by tandem mass spectrometry. In *Agelenopsis aperta* 68 acylpolyamine toxins (Chesnov *et al.*, 2001; Tzouros *et al.*, 2005) and in *Nephilengys borbonica*, 42 acylpolyamine toxins (Itagaki *et al.*, 1997) have been identified. This high diversity of compounds fulfils the principles of combinatorial chemistry to optimise binding to a high diversity of targets of different preys. In the here analysed database, a total of 176 data sets on acylpolyamines are available from 20 species belonging to eight families.

3.3.1 *Amino acid-containing acylpolyamines*

Eighty-two amino acid-containing acylpolyamine toxins occur only in the orb weaver families Araneidae (18 from six *Araneus* and *Argiope* species) and Nephilidae (64 from three *Nephila* and two *Nephilengys* species). From the venom of *Nephilengys borbonica*, 42 acylpolyamine toxins were identified and only five were already known from *Nephila* species (Itagaki *et al.*, 1997). In the venom of the sister species *Nephilengys cruentata*, 25 acylpolyamine toxins were detected; 21 were already known from the venom of *Nephilengys borbonica* and 4 were new (Palma *et al.*, 1998). From the venom of *Nephila clavata* 17 acylpolyamines are described (McCormick and Meinwald, 1993). These few data indicate that each orb-weaver species may have high numbers of compounds in its venom but a given species also shares several compounds with related species. However, both families are separated, since no acylpolyamine toxin occurs in both families.

Today, Araneidae and Nephilidae are regarded as two separate but closely related families. In the past, nephilids were frequently moved between Araneidae and Tetragnathidae and only recently it has been widely accepted that their best supported position is the status of an independent family at the basis of a family group (Araneoidea) comprising also Araneidae, Tetragnathidae, Linyphiidae, Theridiidae and seven other small families (Kuntner *et al.*, 2008). In this context it is important to analyse how different acylpolyamines of Araneidae and Nephilidae are. Taking the classification system of Itagaki and Nakajima (2000), acylpolyamines of Nephilidae belong mostly to structural types A to E, whereas the acylpolyamines of Araneidae belong to structural type F and other types. The molecular masses are rather similar (Araneidae molecular mass range 373–759 Da, mean 620 ± 106 Da; Nephilidae molecular mass range 466–957 Da, mean 683 ± 140 Da; Fig. 5).

3.3.2 *Non-amino acid-containing acylpolyamines*

Non-amino acid-containing acylpolyamine toxins are identified in six families of non-orb weaving spiders, namely three mygalomorph families (Ctenizidae, Hexathelidae and Theraphosidae) and three other families (Agelenidae,

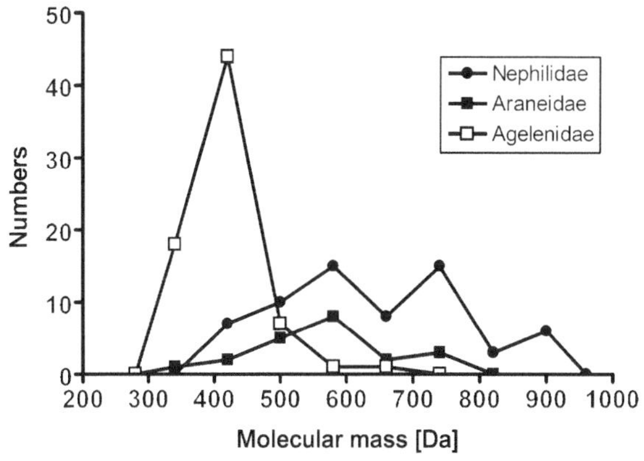

FIG. 5 Molecular mass distribution of amino acid-containing acylpolyamines (Nephilidae, Araneidae) and non-amino acid-containing acylpolyamines (Agelenidae) in three spider families (see Section 3.3).

Amaurobiidae and Pisauridae). These acylpolyamines are on average 150 Da smaller that the amino acid-containing acylpolyamine toxins (molecular mass range 359–728 Da, mean 463 ± 74 Da; Fig. 5). Several agelenid species may share the same compounds, thus the situation in spider families possessing non-amino acid-containing acylpolyamines is similar to the situation in Araneidae or Nephilidae (with amino acid-containing acylpolyamines) where related species also may have the same compounds.

Phylogenetic patterns are difficult to detect since mygalomorphs and the other spider families are at distant ends in the spider system and since the three mentioned entelegyne families are not closer related (Fig. 2). In addition, all families possessing non-amino acid-containing acylpolyamine toxins primarily rely on other venom components, mainly mini-proteins, whereas Araneidae and Nephilidae predominantly rely on the toxic effect of their acylpolyamine toxins. From this lack of records it becomes obvious that non-amino acid-containing acylpolyamine toxins are probably much more wide-spread than it is known so far.

3.4 LINEAR PEPTIDES

Besides disulphide containing peptides, linear peptides of diverse length, cationic charge and biological functions have been identified in the venom of spiders, scorpions and other arthropods (Kuhn-Nentwig, 2003, 2009). These peptides can further be divided into small cationic peptides (< 6 kDa), able to adopt an α-helical conformation (α-SCP) or not (SCP), and a second group of large α-helical cationic peptides (α-LCP) with molecular masses > 7 kDa have been described (Vassilevski *et al.*, 2008). In the past years, the focus was orientated towards the identification of biological active compounds, such as hormone-like peptides and cytolytically acting peptides, which could be useful in drug development (Pimenta and De Lima, 2005).

3.4.1 *Small cationic peptides*

Up to now, SCPs are reported only from *Phoneutria nigriventer* (14 records). Of interests are bradykinin-potentiating or bradykinin-related peptides inhibiting angiotensin-converting enzyme (ACE), which have first been discovered in snake venom (Ferreira and Rocha-e-Silva, 1965) and subsequently identified in the venom of several other venomous organisms including wasps (Mendes and Palma, 2006; Picolo *et al.*, 2010), scorpions (Verano-Braga *et al.*, 2008) and spiders (Ferreira *et al.*, 1998; Sosnina *et al.*, 1990). These peptides are able to act as hypotensive agents in vertebrates and may inhibit insect ACE homologues (Bingham *et al.*, 2006; Lamango *et al.*, 1996). Fortunately, due to peptide micro-scale analytical techniques more and more small linear peptides have been detected in very low concentrations in spider venom. Pimenta *et al.* (2005) identified a family of small linear peptides (PnTkPI-XV: *Phoneutria nigriventer* tachykinin peptides) with molecular masses between 871 and 1654 Da and carrying posttranslational modifications. These peptides exhibit a high similarity with tachykinins, a peptide family characterised by vasodilatory and neurohormonal activities. As mentioned by Piek (1991), kinins can also act as neurotoxic compounds blocking pre-synaptically the cholinergic transmission in the insect central nervous system. Interestingly, multiple bradykinin-related peptides also have been isolated from the sticky droplets in the orb web of *Nephila clavipes* (Volsi *et al.*, 2006) that pointed to an involvement of these peptides into prey immobilisation.

3.4.2 *Small and large α-helical cationic peptides*

Most of the so far identified linear peptides from spider venom concern α-SCP which are in different scales able to destroy prokaryotic and/or eukaryotic cell membranes (Kuhn-Nentwig, 2003, 2009). Responding to the investigated target membranes they have also been named cytolytic, membranolytic or antimicrobial peptides, when apparently ''selectively'' killing bacteria (Matsuzaki, 2009). These peptides usually do not contain cysteine residues and have a higher amount of lysines and arginines, thus they have relatively high positive net charges between + 3 for lycocitin1 (M-lycotoxin-Ls3a; Budnik *et al.*, 2004), + 10 for Oxki1 (M-oxotoxin-Ot1a; Corzo *et al.*, 2002) and at neutral pH pI's > 10. The length of the spider venom α-SCPs identified so far varies between 18 and 48 amino acid residues (Budnik *et al.*, 2004; Corzo *et al.*, 2002). They are disordered in aqueous solutions, but adopt an α-helical structure in the presence of negatively charged membranes. These basic peptides are attracted to the cell surfaces by electrostatic interactions between their positively charged side chains of lysine and arginine and negatively charged membrane phospholipid headgroups, phosphate groups containing lipopolysaccharides or teichoic acids of cellular membranes.

α-SCPs are composed of more than 37% hydrophobic amino acids. The required amphipathicity for their membrane destructing activity is given by spatial distinct hydrophobic and hydrophilic patches on the peptide surface (Powers and Hancock, 2003). Different membrane disrupting models have been discussed: in the carpet model, the cytolytic peptides accumulate in high concentrations parallel to the membrane surface and orientate themselves with their hydrophilic patches to the polar head groups of the phospholipids whereas their hydrophobic patches are faced towards the membrane lipids. After a critical threshold is maintained, development of transient toroidal pores is proposed, followed by destruction of the bilayer curvature and disrupting of the membrane in a detergent-like manner (Shai, 2002). In the barrel-stave model, α-SCPs form the pore in which their hydrophilic patches are orientated to the interior of the pore and the hydrophobic patches face the lipid core region as proposed for alamethicin (Brogden, 2005). In contrast, in the toroidal-pore model the peptides are perpendicularly inserted into the membrane, and peptides are associated with the lipid headgroups lining the pore (Matsuzaki, 1998).

α-SCPs from different arthropod venoms have to be seen in context of antimicrobial peptides identified from the innate immune system of invertebrates and vertebrates (Bulet *et al.*, 2004). More than 1673 peptides are presently specified in two databases (http://www.bbcm.univ.trieste.it/~tossi/amsdb.html and http://aps.unmc.edu/AP/main.php; Wang and Wang, 2004). According to their secondary structures, antimicrobial peptides can be categorized in three main groups: (1) short α-helical peptides (2) short cysteine containing peptides, forming disulphide bridges and (3) extended peptide chains with an overrepresentation of single amino acids. In spider hemocytes, bactericidal peptides belonging to groups 2 and 3 have been identified (Baumann *et al.*, 2010a,b; Lorenzini *et al.*, 2003; Silva *et al.*, 2000), but up to now no α-SCPs exhibiting antimicrobial activity has been described. This may indicate that α-SCPs such as cupiennins in spider venom primarily evolved in the context of a broad cytolytic activity rather than as tools against bacteria.

Interestingly, Vassilevski *et al.* (2008) reported for the zodariid *Lachesana tarabaevi* both α-SCPs (latarcins) and a group of large cationic peptides (α-LCPs), composed of 69–75 amino acid residues, a C-terminal free acid, a net charge of + 14 and pI's ~ 10.2. CD-spectroscopy of CIT1a (M-zodatoxin-Lt8a) exhibits two α-helical regions from 1–34 and 39–69. These helical domains are connected by a short sequence of EEAQ or EEAE and it is supposed that two short peptides form the large peptide in a "head-to tail" orientation. Their antimicrobial and cytolytic activity is comparable to the effects of the small peptides. Generally, most α-SCPs act with sub- to micro-molar range cytolytic activity on a variety of negatively charged cell types of pro- and eukaryotic origin in which, beside the membrane potential, the membrane lipid composition determines their efficiency (Corzo *et al.*, 2002; Giangaspero *et al.*, 2001; Kozlov *et al.*, 2006; Kuhn-Nentwig *et al.*, 2011; Tossi and Sandri, 2002).

The molecular mass range of α-SCPs and α-LCPs varies from 1959 to 8571 Da, which corresponds to 18–75 amino acids. In total there are 88 records, concerning eight families and 17 species. However, these records are very unevenly distributed among taxa: 36 records concern *Cupiennius salei* (Ctenidae), 28 records *Lachesana tarabaevi* (Zodariidae) and the remaining 24 records cover six families with 15 species (Table 5). Also the quality of information is different. For *Cupiennius salei*, *Lachesana tarabaevi* and 14 further data sets, the amino acid sequence of the cytolytic peptides is known. In the case of 10 other species, this is not the case and the information on cytolytic activity is based only on whole venom effects on various cultured cells and bacteria (Benli and Yigit, 2008; Cohen and Quistad, 1998). This could be due to cytotoxic peptides but also to side effects of other venom components. This restriction refers to (1) six species from Agelenidae, Araneidae, Salticidae and Theridiidae, where no other reference to cytolytic peptides has been published so far, thus a side effect is most probably, and (2) three further species

TABLE 5
Spiders which possess cytolytic peptides (α-SCPs and α-LCPs) in their venom

Family	Genus	Species	Evidence		
Agelenidae	*Agelena*	*labyrinthica*			Weak
Araneidae	*Eriophora*	*edax*			Weak
	Neoscona	*arabesca*			Weak
Ctenidae	*Cupiennius*	*salei*	Strong		
Lycosidae	*Geolycosa*	*riograndae*	Strong		
	Hogna	*carolinensis*	Strong		
	Hogna	*helluo*		Probable	
	Lycosa	*erythrognatha*	Strong		
	Lycosa	*singoriensis*	Strong		
	Trochosa	sp.		Probable	
Oxyopidae	*Oxyopes*	*takobius*	Strong		
	Peucetia	*viridans*		Probable	
Salticidae	*Phidippus*	*ardens*			Weak
	Phidippus	*johnsoni*			Weak
	Phidippus	*octopunctatus*			Weak
Theridiidae	*Latrodectus*	*hesperus*			Weak
Zodariidae	*Lachesana*	*tarabaevi*	Strong		

The evidence for the existence of cytolytic peptides is *strong* (compound isolated, sequence known, tests performed or plausible due to sequence homology), *probable* (no compound isolated but cytolytic peptides are known from the same family) or *weak* (sequence not isolated, tests include only crude venom in cell culture) from Lycosidae and Oxyopidae, where further information on cytolytic peptides is available from other species, thus their existence is highly probable.

Following this argumentation, cytolytic peptides are to date only known from two spider groups, Lycosoidea [Ctenidae (Kuhn-Nentwig *et al.*, 2002b), Lycosidae (Budnik *et al.*, 2004; Santos *et al.*, 2010; Yan and Adams, 1998) and Oxyopidae (Corzo *et al.*, 2002)] and Zodariidae (Kozlov *et al.*, 2006; Vassilevski *et al.*, 2008). Both are closer related taxa since the one is the sister group of the other but they differ in the structure of their cytolytic peptides. In Lycosoidea the molecular mass range varies from 1910 to 5221 Da, whereas cytolytic peptides in Zodariidae consist of two groups, one with molecular masses between 2425 and 4341 Da (α-SCPs, also referred to as latarcins), and one with molecular masses between 7866 and 8571 Da (α-LCPs, also called cyto-insectotoxins; Fig. 6).

Within Lycosoidea, ctenid and lycosid cytolytic peptides are rather similar, whereas oxyopid cytolytic peptides possess on average 13 more amino acid residues. The positive charge is mainly given by lysine residues, but in some SCPs from *Cupiennius salei* with unknown function up to 50% arginines appears (Table 6). Cytolytically acting α-SCPs of Lycosoidea and Zodariidae never contain cysteine, but two of the α-LCPs cyto-insectotoxins from *Lachesana tarabaevi* contain one cysteine, which is rather unusual and may confer specific reactivity of the free thiol group.

Interestingly, the N-terminal amino acid residue for the so far known cytolytic peptides from spider venom is mainly glycine, followed by phenylalanine, lysine, serine, alanine, glutamine and valine. A C-terminal amidation is more common than a free C-terminus. Up to now, no clear functional explanation for the C-terminal amidation of antimicrobial peptides could be given (Dennison *et al.*, 2009). They demonstrated for short naturally and synthetic defence peptides that C-terminal amidation has variable effects on the antimicrobial and anticancer activity, but without clear effects on the selectivity towards these cell types. Investigations of cupiennin 1a (M-ctenitoxin-Cs1a) with the amidated form or with a free C-terminus exhibited no differences in their biological activity (Kuhn-Nentwig *et al.*, 2002b).

Within a peptide, glycine and proline are known to induce a helix–hinge–helix structure in α-helical peptides resulting in a flexible disordered area enclosed by helices as shown for cupiennin 1a (M-ctenitoxin-Cs1a; Gunasekaran *et al.*, 1998; Pukala *et al.*, 2007a; Woolfson and Williams, 1990). In such a function, glycine is most dominant, but 3 out of 12 *Lachesana tarabaevi* α-SCPs (latarcins) contain one proline, also 7 out of 33 peptides from *Cupiennius salei* (Trachsel, 2010) and one peptide from *Oxyopes takobius*.

The main cytolytic acting peptides isolated from *Cupiennius salei* comprises the cupiennin 1 and cupiennin 2 families, which are characterised by a hydrophobic N-terminus connected to the C-terminus via several lysine repeat segments of four residues. The assembly of these repeats can be found in peptides from *Lycosa erythognatha* and *L. singoriensis*, *Lachesana tarabaevi* and

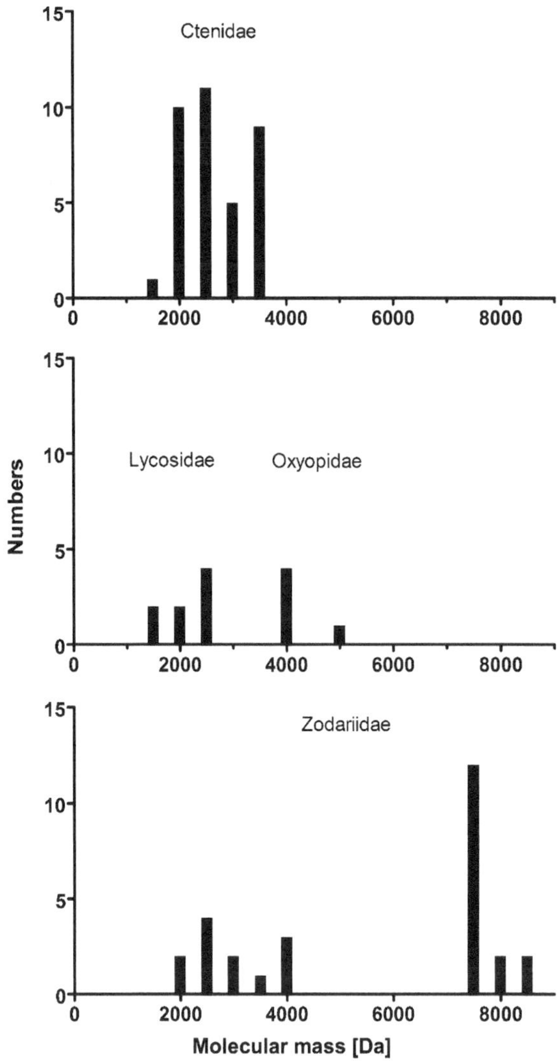

FIG. 6 Molecular mass distribution of cytolytic peptides in Lycosoidea (Ctenidae (*Cupiennius salei*), Lycosidae, Oxyopidae) and Zodariidae (*Lachesana tarabaevi*).

Oxyopes takobius. Moreover, several peptides exhibit a hydrophobic N-terminus with a distinct amphipathic conformation, while the C-terminus is more hydrophilic (Fig. 7). Investigations with cupiennin 1d (M-ctenitoxin-Cs1d) demonstrated that the N-terminus determines the cytolytic activity whereas the C-terminus modulates the accumulation at negatively charged membranes (Kuhn-Nentwig *et al.*, 2002a).

TABLE 6
Characteristics of cytolytic peptides from four spider families

Family	Number of cytolytic peptides	Average molecular mass (Da ± SD)	Average number of Amino acids	Average number of Lysines	Average number of Arginines
Ctenidae	36	2875 ± 605	25.9 ± 6.6	6.1 ± 1.8	0.5 ± 0.9
Lycosidae	9	2809 ± 501	24.3 ± 3.7	6.0 ± 2.6	0.2 ± 0.4
Oxyopidae	5	4343 ± 492	39.2 ± 4.9	8.4 ± 0.5	2.0 ± 0.7
Zodariidae SCPs	12	3303 ± 691	27.5 ± 5.5	6.5 ± 1.2	1.6 ± 1.2
Zodariidae LCPs	16	8031 ± 245	70.0 ± 2.1	19.0 ± 0.5	1.3 ± 0.4

Ctenidae, Lycosidae and Oxyopidae constitute the Lycosoidea superfamily which is the sister group to Zodariidae

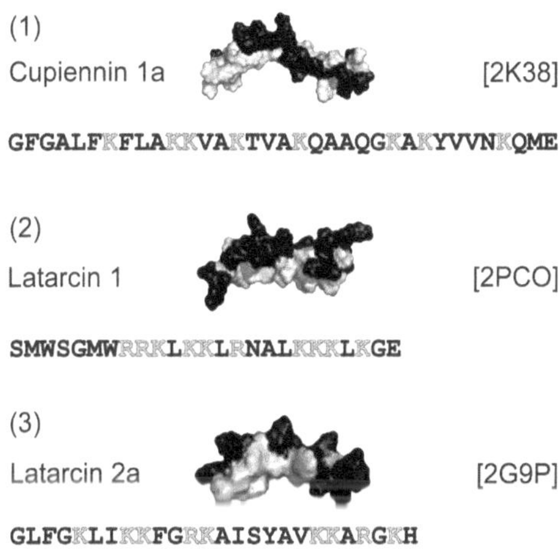

FIG. 7 Amino acid sequences and three-dimensional structures of (1) short cationic α-helical peptides from the ctenid *Cupiennius salei* and (2 and 3) the zodariid *Lachesana tarabaevi*. Positively charged amino acids in the amino acid sequences are given in grey. In the spatial structure of the peptides charged/polar residues are shown in black and hydrophobic amino acids in grey. The Protein Data Bank codes are given in brackets. The structures were calculated using PyMOL (DeLano, 2002).

Cytolytic peptides occur in spider venom alone or in different combinations together with low molecular mass compounds, neurotoxins and enzymes (Corzo *et al.*, 2002; Kuhn-Nentwig *et al.*, 2002b; Villegas *et al.*, 2008). In *Lachesana tarabaevi*, beside the small latarcins (Kozlov *et al.*, 2006) also unusual cyto-insectotoxins were identified (Vassilevski *et al.*, 2008). Today, they probably constitute the largest α-helical cytolytic peptides and it has been argued that they developed from a duplication of small cytolytic peptides. The insecticidal effects of cyto-insectotoxins is superior to latarcins and therefore this modification was a very successful evolutionary step for zodariids, if this can be generalised from the one investigated species to the whole family (Vassilevski *et al.*, 2009). Nothing is known about low molecular mass compounds and mini-proteins, which may be present in the venom and may synergistically support these peptides. In opposite, in the venom of the lycosoid species, beside cytolytic peptides (Budnik *et al.*, 2004; Santos *et al.*, 2010; Yan and Adams, 1998), several low molecular mass compounds (Margaret and Phanuel, 1988; Schroeder *et al.*, 2008), mini-proteins (Cruz *et al.*, 1994; Fisyunov *et al.*, 2005; Zhang *et al.*, 2010) and enzymes (Devaraja *et al.*, 2008; Nagaraju *et al.*, 2006, 2007a,b) have been identified (see Section 3.7.2).

Functionally speaking, cytolytic peptides can act as toxins, but they will only paralyse or kill a victim if they are produced in sufficient quantity, which is the case in *Lachesana tarabaevi*. *Zodarion* species are known as ''ant-specialists'' (Cushing and Santangelo, 2008) but *L. tarabaevi* obviously feeds on ants, as well as on woodlice according to limited field observations (Zonstein and Ovtchinnikov, 1999). Further ecological studies and venom investigations in more zodariid species would help to understand the development of cyto-insectotoxins in terms of prey specialisation.

A completely different function of cytolytic peptides is a synergistic mode of action to enhance the effect of mini-proteins. This has been shown for cupiennins (Kuhn-Nentwig *et al.*, 2004; Wullschleger *et al.*, 2005) and oxyopinins (Corzo *et al.*, 2002). Besides their cytolytic activity, α-SCPs such as cupiennin 1a (M-ctenitoxin-Cs1a) are able to inhibit the formation of nitric oxide by neuronal nitric oxide synthase (nNOS; Pukala *et al.*, 2007b), as well as the production of superoxide by NADPH oxidase in phorbol myristate acetate stimulated granulocytes (Kuhn-Nentwig *et al.*, 2011).

Other authors have mentioned three further functions of cytolytic peptides. They assume that these peptides could also be used as protection of the venom gland against microbial infections (Conde *et al.*, 2000; Dai *et al.*, 2002; Moerman *et al.*, 2002; Yan and Adams, 1998) and as antiseptics (Orivel *et al.*, 2001; Torres-Larios *et al.*, 2000). While such purposes cannot be excluded, spiders usually use digestive fluid to clean themselves after a meal. This enzyme mixture is certainly highly effective against microbes, additionally it is available in larger quantities and possibly also energetically easier to produce than venom components. As third possibility, a digestive function has been mentioned by Vassilevski *et al.* (2009). Indeed, cytolytic peptides exert some cell and tissue destroying effects and certainly support digestion, but in spiders envenomation and digestion are two clearly separated processes with digestive fluid being the liquid of choice to perform digestion as efficiently as possible (see Section 4.5).

3.5 CYSTEINE-KNOTTED MINI-PROTEINS

A total of 965 mini-proteins are known so far from the venom of 60 spider species belonging to 20 families. This concerns 529 records for mygalomorph spiders (five families), 41 records for haplogyne spiders (five families), 162 for Lycosidae, 114 for Ctenidae, 70 for Agelenidae and 49 for seven further families. The increasing number of proteomic, peptidomic and transcriptomic studies indicates that many mini-proteins occur in one venom and more than 100 different toxins or variations of a few different toxin types are not uncommon in a single venom (Chen *et al.*, 2008a,b; Diego-Garcia *et al.*, 2010; Fernandes-Pedrosa *et al.*, 2008; Gremski *et al.*, 2010; Jiang *et al.*, 2010; Liang, 2008; Liao *et al.*, 2007; Tang *et al.*, 2010; Yuan *et al.*, 2007a; Zhang *et al.*, 2010).

3.5.1 *Functional aspects*

Mini-proteins comprise peptides with molecular masses between 2650 and 14,800 Da with most toxins between 3000 and 9000 Da. They consist of 6–14 cysteines and exert a typical complex pattern of disulphide bridges. Mainly, three structural motif types in spider polypeptide chain folds occur:

(1) The ''inhibitor cysteine knot'' (ICK) motif exhibiting a spatial structure consisting of a β-hairpin and a ''knot'' built by the C^3—C^6 bond penetrating the ring formed by the two other bonds and the involved amino acids (Norton and Pallaghy, 1998). The disulphide bridges are arranged as C^1—C^4, C^2—C^5, C^3—C^6 as shown for HWTX-IV (μ-theraphotoxin-Hh2a) from *Haplopelma schmidti* with a two-stranded antiparallel β-sheet fold (Peng *et al.*, 2002). A further ICK arrangement is known as C^1—C^4, C^2—C^5, C^3—C^8, C^6—C^7 in which the fourth disulphide bridge C^6—C^7 is introduced into the extended β-hairpin structure as verified for ACTX-Hi:OB4219 with a triple-stranded antiparallel beta sheet fold (Rosengren *et al.*, 2002). In J-ACTX Hv1c, the disulphide bridges are arranged as C^1—C^6, C^2—C^7, C^5—C^8 with a rare vicinal disulphide bridge between Cys13 and Cys14 (C^3—C^4) that is important for its biological activity but not for the stabilization of the peptide (Fig. 8; Gunning *et al.*, 2008; Wang *et al.*, 2000). Beyond spiders, the ICK motif has been detected in many peptides of various functions in fungi, plants and invertebrates as summarized in the KNOTTIN database (Bulet and Stöcklin, 2005; Gelly *et al.*, 2004; Gracy *et al.*, 2008).

Kozlov and Grishin (2005) introduced the ''principal structural motif'' for the structural characterisation of the six half-cystine residues in spider neurotoxins following the ICK motif. This could be helpful for the identification of toxin-like structures in spider venom. The principal structural motif is located in the N-terminal part of the peptides and is defined mainly by six amino acids between Cys1 and Cys2 followed by a variable number of amino acid residues (U) to the occurrence of two adjacent cysteines: C^1-6-C^2-U-C^3C^4. In addition, if more than six half-cystines are present in an ICK containing neurotoxin without

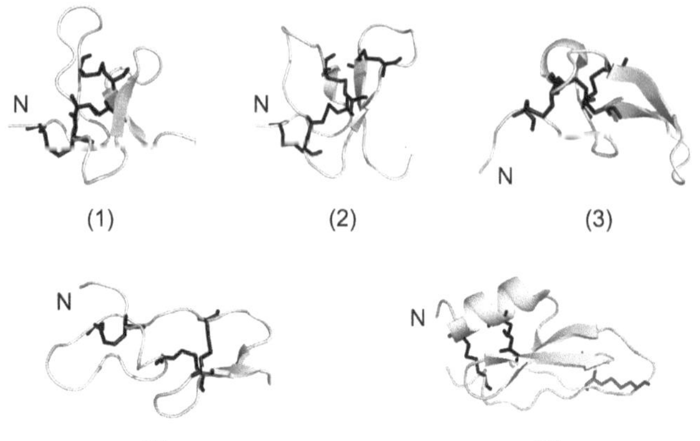

FIG. 8 Spatial structure models of different mini-proteins sharing the ICK motif (1) Huwentoxin-IV (μ-theraphotoxin-Hh2a; PDB code 1MB6) from *Haplopelma schmidti* with six cysteines, (2) ACTX-HI:OB4219 (U_2-hexatoxin-Hi1a; 1KQH) from *Hadronyche infensa* and (3) ω-atracotoxin-Hv1 (ω-hexatoxin-Hv1a; 1AXH) with eight cysteines from *Hadronyche versuta.* The DDH motif is identified in (4) huwentoxin-II (U_1-theraphotoxin-Hh1a; 1I25) from *Haplopelma schmidti* with six cysteines. The Kunitz motif (5) is represented by huwentoxin-XI (κ-theraphotoxin-Hh1a; 2JOT) from *Haplopelma schmidti* with six cysteines. The models were calculated using PyMOL (DeLano, 2002) and cysteines are coloured in black. The N-terminus is marked with N.

altering the ICK motif, an ''extra structural motif'' is defined as C^5-1-C^6-U-C^7-1-C^8. Kozlov *et al.* (2005) describe the ''principle structural motif'' as reliable indicator of a neurotoxic function. Moreover, most of these peptides uniformly have no amino acid between C^3 and C^4 (Table 7). Such a structure obviously makes these folded polypeptides exceptionally stable and suitable as potent and selective toxins.

(2) Another structural motif is given by the ''disulphide-directed β-hairpin'' (DDH) that is supposed to be the ancestral motif and is characterised by an antiparallel β-hairpin stabilized by two mandatory disulphide bonds as shown in HWTX-II (= U1-theraphotoxin-Hh1a) from *Haplopelma schmidti* (Fig. 8). The disulphide arrangement is given as C^1—C^3, C^2—C^5, C^4—C^6 (Shu *et al.*, 2002; Shu and Liang, 1999).

(3) The ''Kunitz motif'' is characterised by a short N-terminal 3_{10} helix, C-terminal α-helix and a three stranded antiparallel β-structure with the following disulphide bridge arrangement: C^1—C^6, C^2—C^4, C^3—C^5. So far, this motif has only been identified in mini-proteins from the venom of two theraphosid spiders, *Haplopelma schmidti*, HWTX-XI (κ-theraphotoxin-Hh1a) exhibiting a potassium channel blocking activity and acting as protease inhibitor (Fig. 8; Peng *et al.*, 2006), and Haplopelma *hainanum* (Tang *et al.*, 2010; Yuan *et al.*, 2008). Nevertheless, this motif is well known from a diversity of peptides and proteins such as bovine pancreatic trypsin inhibitors (BPTIs; Berndt *et al.*,

TABLE 7

Structure of mini-proteins with six cysteines (C1–C6) and no amino acid between C3 and C4

	Number of mini-proteins	Molecular mass (Da)	Number of amino acids: N-terminal	C1–C2	C2–C3	C4–C5	C5–C6	C-terminal	Total
Hexathelidae	61	4124	2.6	5.9	5.0	4.1	8.5	7.4	33.4
Theraphosidae	122	4017	1.3	6.0	5.5	4.2	6.3	5.5	28.9
Agelenidae	25	3997	2.0	6.8	6.5	3.9	8.9	1.5	30.2
Sparassidae	7	3705	1.9	6.0	5.1	4.0	4.9	4.4	27.7
Ctenidae	14	3903	1.9	5.9	5.1	4.4	7.8	3.4	26.3

Average molecular masses, number of amino acids between neighbouring cysteines and total number of cysteines are given for five spider families.

1992), neurotoxins of snakes (Strydom, 1973), ticks (Sasaki *et al.*, 2004), cone snails (Dy *et al.*, 2006) and sea anemones (Schweitz *et al.*, 1995).

Mini-proteins act selectively against a specific target and form a tight and stable toxin-target complex inducing prolonged bioactivity. They mainly act on membrane proteins in electro excitable cell membranes, that is neuronal and muscular cells, primarily modulating ion channels such as calcium (Ca^{2+}), sodium (Na^+) and potassium (K^+), but also on mechano-, chemo-, thermo-sensitive receptors, and inhibit, activate or delay voltage-activated channels so that their normal ion regulation is affected (Table 8). Usually, these toxins are very effective already at nano-molar concentrations, which is at least one order of magnitude better than the average concentration needed for the unspecific membrane destruction by cytolytic peptides, for example, when comparing CSTX-1 (ω-ctenitoxin-Cs1a), a Ca^{2+} L-type channel blocker (Kubista *et al.*, 2007) and cupiennin 1a (= M-ctenitoxin-Cs1a; Kuhn-Nentwig *et al.*, 2002b), both identified in the venom of *Cupiennius salei*.

Spider venom mini-proteins are usually made of the common 20 amino acids, and C-terminal amidation is not uncommon. Only a few unusual amino acids are known so far, such as the O-palmitoyl threonine amide residue, modifying the C-terminus of PLTX-II (ω-plectoxin-Pt1a) from the plectreurid *Plectreurys tristis*, which is essential for toxin activity (Branton *et al.*, 1993); or the altered chirality of Ser46 to the D-configuration in ω-agatoxin IVB (ω-agatoxin-Aa4b) from the agelenid *Agelenopsis aperta* that improves the activity of the toxin nearly by two orders of magnitude (Heck *et al.*, 1994).

TABLE 8

Activity descriptors and biological function of mini-proteins, nomenclature according to King *et al.* (2008), and the number of known occurrences in spiders

Target ion channel	Descriptor	Function	Number of mini-proteins
Ca^{2+}	ω (omega)	Inhibits voltage-activated calcium (Ca_v) channels	111
K^+	κ (kappa)	Inhibits voltage-activated potassium (K_v) channels	75
Na^+	β (beta)	Shifts voltage dependence of Na_v channels activation	48
	δ (delta)	Delays inactivation of voltage-activated Na_v channels	50
	μ (mu)	Inhibits voltage-activated sodium (Na_v) channels	37
others	γ (gamma)	Targets hyperpolarisation-activated, cyclic nucleotide-gated nonspecific cation channels	1
	π (pi)	Targets acid-sensing ion channels	1
	τ (tau)	Targets transient receptor potential channel	3

There are hundreds of primary references dealing with detailed descriptions of the activity of nearly 1000 toxins from the venom of 60 spider species on a large number of arthropod and vertebrate targets. Biological activities are typically investigated *in vivo* in model insects and by techniques such as binding studies or electrophysiology, to investigate potency and selectivity for selected sub-types of ion channels, receptors or other targets. These variations in response are highly interesting to develop novel pharmacological tools or to discover new lead compounds from a pharmaceutical point of view where specific inhibitors of specialized receptor types or ion channel types are wanted to develop novel drugs. Reviews on the latter aspects have been provided by several research groups (Borges *et al.*, 2009; Bosmans *et al.*, 2009; Escoubas and Rash, 2004; Escoubas *et al.*, 2000b; Estrada *et al.*, 2007; Lewis and Garcia, 2003; Rash and Hodgson, 2002; Stöcklin and Vorherr, 2010; Vassilevski *et al.*, 2009). Of course, from a generalized functional point of view and from the evolutionary sight of venom development within spiders, pharmaceutical applications are not the driving force.

3.5.1.1. Calcium channel modulators In most tissues of animals, especially in the nervous system and in muscles, calcium channels regulate the release of neurotransmitters. They can be inhibited by toxins causing a long-lasting specific blockade of these pre-synaptic voltage-activated calcium (Ca_v) channels. There is a confusing historic nomenclature of such toxins and today they are called omega-toxins (ω-toxins). There are six recognised types and many sub-types of Ca_v channels (L, N, P, Q, R and T) differing in their electrophysiological characteristics and they react to different inhibitors or activators differently (Rash and Hodgson, 2002).

From spider venoms, so far 111 ω-toxins have been identified (corresponding roughly to 50% of all functionally characterized mini-proteins) and they differ considerably in molecular masses (2938–8752 Da, Fig. 9) and among families: in the mygalomorphs Actinopodidae, Hexathelidae and Theraphosidae ω-toxins always contain six cysteines (Chong *et al.*, 2007). In the haplogynes Filistatidae, Plectreuridae and Segestriidae ω-toxins contain 10–12 cysteines. In Agelenidae they consist of 8–12 cysteines and in Ctenidae 6–12 cysteines, whereas in further entelegyne families, cysteine numbers are fixed (Sparassidae only with 6 cysteines, Lycosidae with 8 cysteines, Oxyopidae with 10 cysteines).

Typical calcium channel toxins are ω-agatoxins from the agelenid *Agelenopsis aperta* with subtype I selectivity for L-type channels, subtype II selectivity for N-type channels, or subtype IV selectivity for P/Q-type channels (Adams *et al.*, 1990, 1993; Bindokas and Adams, 1989; Venema *et al.*, 1992). Further examples concern ω-GsTx-SIS (ω-theraphotoxin-Gr1a) from the theraphosid *Grammostola spatula* (Piser *et al.*, 1995) and Tx3-3 or Tx3-4 (ω-ctenitoxin-Pn2a and ω-ctenitoxin-Pn3a) from the ctenid *Phoneutria nigriventer* selectively acting on P/Q-type Ca_v channels (Cordeiro *et al.*, 1993; Miranda *et al.*, 1998).

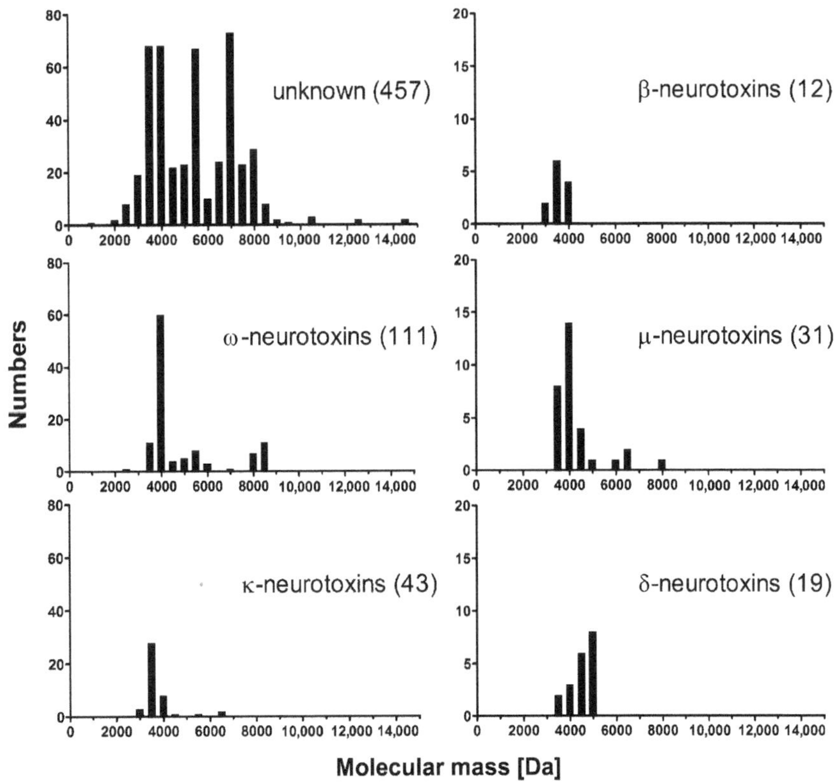

FIG. 9 Molecular mass distribution of mini-proteins/cysteine-knotted peptides from spider venom with respect to their function as ω-, κ-, β, μ- and δ-toxins. For most toxins, however, the exact mode of action remains unknown. Numbers in brackets refer to the number of records.

3.5.1.2. Sodium channel modulators The fast transport of impulses along and between excitable cells is caused by the movement of sodium (Na^+) ions across the membranes of excitable cells via voltage-activated sodium (Na_v) channels. These ion channels enable the coordination of locomotion and for this reason they are usually the most abundant ion channels in nerve and muscle tissue. However, importance and frequency makes sodium channels also the first target for paralytic neurotoxins and some of the Na_v sub-types represent a validated pharmacological target to develop new pain killing drugs. There is a high diversity in this family of sodium channels made of at least nine members: Na_v1.1–Na_v1.9 (Goldin, 1999).

Depending on the precise mode of action, three toxin types are distinguished (Table 8). Spider beta-toxins (β-toxins) shift the voltage dependence of Na_v channels activation, their molecular masses range from 3292 to 4331 Da and they contain six cysteines. From spider venoms, so far 48 β-toxins are known from five theraphosid, two hexathelid and one agelenid species. β-toxins are known, for example, from the theraphosid *Ceratogyrus marshalli* (Ceratotoxin-1 = β-theraphotoxin-Cm1a; Bosmans and Swartz, 2010; Bosmans *et al.*, 2006).

Spider delta-toxins (δ-toxins) delay the inactivation of Na_v channels and range from 3681 to 5298 Da. Fifty δ-toxins identified from the venom of one actinopodid, three hexathelid species, one theraphosid (containing six to eight cysteines), one amaurobiid (eight cysteines), one agelenid (eight cysteines) and three ctenid species (10–11 cysteines). δ-toxins are robustoxin (RTX = δ-hexatoxin-Ar1a) and versutoxin (VTX = δ-hexatoxin-Hv1a) from the Australian funnel-web spider *Atrax robustus* and from *Hadronyche versuta* (both Hexathelidae; Borges *et al.*, 2009; Nicholson *et al.*, 2004, 2006).

Spider mu-toxins (μ-toxins) inhibit voltage-activated sodium (Na_v) channels and their molecular masses vary from 3612 to 8612 Da. The known 37 μ-toxins derive from one hexathelid species (6–10 cysteines), four theraphosid species (six cysteines), one diguetid species (eight cysteines), three agelenid species (seven to eight cysteines), one thomisid (six cysteines) and one ctenid species (14 cysteines). μ-toxins are known as μ-agatoxins (μ-Aga-I to IV) from *Agelenopsis aperta* or as curtatoxins from *Hololena curta* (both Agelenidae; Adams, 2004; King *et al.*, 2008; Nicholson, 2007).

3.5.1.3. Potassium channel modulators Potassium channels comprise a highly diverse and large group of ion channels, regulating different cellular processes in the body. Voltage-activated potassium (K_v) channels consist of four main subunits and several accessory subunits to form an ion pore.

Spider kappa-toxins (κ-toxins) inhibit voltage-activated potassium channels and there are 75 records for spider species, eight theraphosid species (mini-proteins with six cysteines), two hadronychid species (eight cysteines), one sparassid (six cysteines) and one ctenid species (eight cysteines). These toxins have a molecular mass range of 3280–6172 Da. κ-toxins include, for example heteropodatoxins (HpTx 1–3, e.g. κ-sparatoxin-Hv1a) from the sparassid *Heteropoda venatoria* or several toxins from the theraphosid *Chilobrachys guangxiensis* (e.g. κ-theraphotoxin-Cj1a; Liao *et al.*, 2006; Mouhat *et al.*, 2008; Sanguinetti *et al.*, 1997; Swartz, 2007; Yuan *et al.*, 2007b).

3.5.1.4. Interactions with other targets Beside the above-mentioned classical ion channel targets of neurotoxins, further targets have been detected by selective screening of a diversity of spider venoms. In the venom of *Psalmopoeus cambridgei*, for example, several bioactive peptides have been identified: Psalmotoxin 1 (PcTx1; π-theraphotoxin-Pc1a) is a potent blocker of ASIC-1, a member of the acid sensitive ion channels (Escoubas *et al.*, 2000a; Qadri *et al.*, 2009). Additionally, two further ICK-peptides, psalmopeotoxins I and II (PcFK1 = U_1-theraphotoxin-Pc1a and PcFK2 = U_2-theraphotoxin-Pc1a) with antimalarial activities were identified. Both peptides are selectively active against the intra-erythrocyte stage of *Plasmodium falciparum*, but do not lyse erythrocytes, they do not act cytotoxically towards nucleated mammalian cells and do not inhibit the neuromuscular function (Choi *et al.*, 2004; Kamolkijkarn *et al.*, 2010; Pimentel *et al.*, 2006). Vanillotoxins (VaTx1–3, e.g. τ-theraphotoxin-Pc1b) activate the noxious heat-sensing capsaicin receptor (TRPV1) causing pain. Interestingly, VaTx1 (= τ/κ-theraphotoxin-Pc1a) also modulates voltage gating of the K_v2.1 channel (Cromer and McIntyre, 2008; Siemens *et al.*, 2006).

The venom of *Grammostola rosea* is the source of mechanosensitive channel ligands such as α-mechanotoxin (GsAF II = κ-theraphotoxin-Gr2c), GsMTx-2 (κ-theraphotoxin-Gr2a) and GsMTx-4 (M-theraphotoxin-Gr1a; Bowman *et al.*, 2007; Oswald *et al.*, 2002; Suchyna *et al.*, 2000). GsMTx-4 exhibits an ellipsoidal shape with a hydrophobic protrusion at the centre. A ring of positively and negatively charged residues is placed around the periphery (Nishizawa and Nishizawa, 2007). This amphipathic peptide acts as bactericidal on $Gram^+$ as well as on $Gram^-$ bacteria (Hurst *et al.*, 2009; Jung *et al.*, 2006). Remarkable are results from Redaelli *et al.* (2010) revealing that GsMTx-4 and GsAF II exhibit target promiscuity and heterogeneous effects on Na_v and K_v channels. As described in detail by Bosmans *et al.* (2009), six further mini-proteins isolated from different mygalomorph venoms exhibit heterogenous effects on Na_v and K_v channels, on Ca_v and K_v channels, or even on Ca_v, K_v and Na_v channels. This supports the assumption that many mini-proteins may show a rather broad or less specific activity (promiscuous), at least when broad and comparative tests are performed.

A lectin-like peptide (SHLP-I = U_5-theraphotoxin-Hh1a) was identified in a venom gland cDNA library of *Haplopelma schmidti* and further produced by recombinant means. Jiang *et al.* (2009) could demonstrate an agglutination of human erythrocytes by the purified synthetic peptides SHL-Ib1b and SHL-Ib1c. From the same spider, a TRPV1 agonist was also isolated (Bohlen *et al.*, 2010). In contrast to the above-mentioned vanillotoxins, the ''double-knot'' toxin (DkTx) is a bivalent ICK peptide (8.5 kDa), composed of two independently folded head-to-tail ICK domains that are connected by a kinked ether. It is supposed that both spiders *Psalmopoeus cambridgei* and *Haplopelma schmidti* developed independently TRPV1 agonists (convergent evolution). A comparable mechanism of head-to-tail doubling of peptides is to observe for the cyto-insectotoxins from *Lachesana tarabaevi* (see Section 3.4.2; Vassilevski *et al.*, 2008) and CpTx1 from *Cheiracanthium punctorium* (Vassilevski *et al.*, 2010), both araneomorph spiders.

3.5.2 *Evolutionary aspects*

3.5.2.1. Replacement of mini-proteins Although we present data on 965 mini-proteins from 60 spider species belonging to 20 families, no information on mini-proteins is available for 12 other families presented in this review, most of which are only sporadically investigated. So, a first conclusion could be that all spiders possess mini-proteins. However, the lack of mini-proteins in a few families is remarkable and probably not due to research gaps: while Nephilidae seem to rely on acylpolyamines (Itagaki *et al.*, 1997; Palma *et al.*, 1998),

Zodariidae rely mainly on cytolytic peptides (Vassilevski *et al.*, 2009) and Theridiidae possess large neurotoxic proteins (Rohou *et al.*, 2007).

The venom of the yellow sac spider *Cheiracanthium punctorium* (Miturgidae) contains a 16-cysteine residues peptide (CpTx1) with a molecular mass of 15.1 kDa, consisting of two distinct primary structures both following the general ICK motif (Vassilevski *et al.*, 2010). This suggests already a function on ion channels, but overall sequence homology with other neuroactive peptides is rather low. Moreover, effects on a number of ion transport proteins could be excluded and the general structure of a mini-protein may rather reflect its developmental origin than its actual function. CpTx1 is a large cationic peptide with a net charge of $+8$ (pH 7, pI $\approx$ 10) forming amphipathic structures, that is revealing hydrophobic and hydrophilic regions that may effectively interact with membranes and cause a pronounced cytolytic effect. As a two-domain modular toxin, CpTx1 is clearly different from all other toxin families. Evolved in a small spider family and embedded in the Lycosoidea family group, this may indicate an evolutionary step away from specific modulatory activities on ion channels towards cytolytic actions at the cytoplasmic membrane lipid bilayers. This would correspond to a development away from specialised toxins or target receptors to a more generalized and less specific effect. This idea is supported by the very limited homology of CpTx1, only found with two other Lycosoidea families, Ctenidae and Lycosidae (Vassilevski *et al.*, 2010).

Vice versa, there is evidence of mini-proteins in spider families where it would not have been expected. Among Sicariidae, there are six records of mini-proteins for *Loxosceles intermedia*, while 72 records concern enzymes and low molecular compounds in many other *Loxosceles* and *Sicarius* species. Probably due to narrowed research focus on phospholipase D (PLD; see below; Binford and Wells, 2003), the nonetheless important family of mini-protein toxins may well have been overlooked (de Castro *et al.*, 2004; Fernandes-Pedrosa *et al.*, 2008).

It has been widely accepted that Araneidae, paralleled by Nephilidae, replaced mini-proteins by acylpolyamines. However, at the transcript level three mini-proteins are known to occur in *Araneus ventricosus* and *Caerostris extrusa* (Herzig *et al.*, 2011) which does not really fit the common assumption that this family exclusively relies on acylpolyamines. The two *Araneus ventricosus* toxins (U_1-aranetoxin-Av1a and U_2-aranetoxin-Av1a) contain a BPTI/ Kunitz-like inhibitor domain and have originally been described as ''insecticidals'' (Jung *et al.*, 2002). However, the structure of these peptides is clearly different from the typical ICK motif and their main function may well be in the field of protease inhibition. The BPTI, also known as pancreatic Kunitz inhibitor, is a globular protein with a broad specificity, inhibiting trypsin-, chymotrypsin- and elastase-like serine proteases by forming stable complexes. Since BPTI also blocks K^+ transport by Ca^{2+}-activated K^+ channels (Ascenzi *et al.*, 2003) and occurs in arthropods (Sasaki and Tanaka, 2008), it is well possible that ion channel blockage is a side effect of protease inhibition. There is certainly more research needed on this topic.

The *Caerostris extrusa* toxin (U_3-aranetoxin-Ce1a) is also difficult to evaluate since there is only evidence at transcript level. The initial peptide has a length of 102 amino acids and the procession to the mature form is expected to yield a 8950-Da peptide composed of 81 amino acids. It contains 10 cysteines and follows the ICK motif (Dai *et al.*, unpublished, cited in Herzig *et al.*, 2011). With the exception of this putative toxin, all Araneidae investigated so far can be counted together with Nephilidae as spider families primarily relying on acylpolyamines.

3.5.2.2. Modifications of mini-proteins Within the mentioned families, mini-proteins are recorded for 60 species, most of which are mygalomorph species (20 species in Theraphosidae, nine in Hexathelidae, one species from Actinopodidae, Cyrtaucheniidae and Nemesiidae each), six species in Ctenidae, five species from Agelenidae, and 13 further families with one to two species each. The species with the highest numbers of recorded mini-proteins (> 30) are *Haplopelma hainanum* (166 mini-proteins; Tang *et al.*, 2010) *Lycosa singoriensis* (154; Zhang *et al.*, 2010), *Chilobrachys guangxiensis* (80; Chen *et al.*, 2008b), *Haplopelma schmidti* (64; Liang, 2004), *Agelena orientalis* (47; Kozlov *et al.*, 2005; Pan *et al.*, 2007), *Phoneutria nigriventer* (45; Borges *et al.*, 2009; Gomez *et al.*, 2002), *Macrothele gigas* (35; Corzo *et al.*, 2003; Satake *et al.*, 2004), and *Hadronyche versuta* (32; Herzig *et al.*, 2011).

In most of these cases, however, the structure of these molecules is only known through expression at the transcript level and further estimated procession into the mature form. The overall structure and especially the pattern of the disulphide knot structure could identify them as peptides belonging to the ICK motif family of spider venom toxins. The function is then assumed according to this structure, the overall similarity in amino acid sequence, a high degree of sequence homology with related peptides having a known function, and its origin or organ of expression: the venom glands. For reasons of simplicity and since no alternative is available, such assumptions tend to be followed. However, there are well-known examples of ICK-peptides that exhibit no obvious toxic activity. In *Cupiennius salei*, for example, the neurotoxic functions of the main toxins CSTX-1 (ω-ctenitoxin-Cs1a) and CSTX-9 (U_1-ctenitoxin-Cs1a) are enhanced by CSTX-13 (U_2-ctenitoxin-Cs1a), which itself has only a negligible own toxicity (Wullschleger *et al.*, 2004, 2005). All three peptides share the common ICK motif and it can be speculated that CSTX-13 developed from an ICK-toxin to an ICK-enhancer with only low own toxicity. In the somewhat related case of *Cheiracanthium punctorium* (Miturgidae) described above, an ICK-precursor peptide evolved into the cytolytic CpTx1 peptide still showing ICK structures (Vassilevski *et al.*, 2010).

In nearly two third of all cases the precise target receptor is not yet known. The 330 records with known target concern 111 ω-toxins (acting on calcium channels), 75 κ-toxins (targeting potassium channels), 37 μ-toxins (modulating sodium channels), 50 δ-toxins and 48 β-toxins affecting sodium channels in different ways, as well as three τ-toxins, one γ-toxin and one π-toxin (Table 8). However, the distribution of these mini-proteins with respect to receptor specificity among spider families is highly speculative, at least as of today. As a result, this present distribution does not necessarily reflect the real situation; it rather refers to research bias with respect to spider taxa or receptor types.

Mini-proteins contain 6–14 cysteines (if we neglect the above-mentioned CpTx1 from *Cheiracanthium punctorium* (Miturgidae) containing 16 cysteines) but the lower and the even cysteine numbers are prevailing. 38% of all toxins contain 6 cysteines, 39% contain 8 and 10% contain 10 cysteines. Toxins with 12 or 14 cysteines are less common (4% and 1%, respectively), and structures with an odd number of cysteines are even less frequent (7 cysteines 4%, 9 cysteines 2%, 11 cysteines 1%) and may be artefactual in some cases. Whenever confirmed, the investigation of the possible role of the highly reactive thiol groups of free cysteines would certainly be of worth efforts. On family level, some taxon-specific peculiarities appear. Theraphosidae mostly rely on peptides with six (77%) and eight cysteines (17%). In Hexathelidae, the situation is similar (69% and 20%, respectively), also in Agelenidae (44% and 39%, respectively). In contrast, Lycosidae mostly possess 8 cysteine peptides (70%), 11% have 10 cysteines and in this family also 12 cysteine and 14 cysteine peptides occur (3% and 4% respectively). This is similar to the ctenid *Cupiennius salei*, which mainly (94%) has peptides with eight cysteines. Opposite to this, the other ctenids show a wide range of cysteine numbers (six cysteines 23%, eight 16%, ten 21%, twelve 13%). All so far known sparassid mini-proteins comprise 6 cysteines, all amaurobiid peptides have 8 cysteines, and all oxyopid peptides have 10 cysteines (Table 9).

The number of lysines and arginines (positively charged residues typically over-represented in antimicrobial and cytolytic peptides) varies from 0 (in peptides from five spider families) to 15 residues (lasiotoxin-1 = U_1-theraphotoxin-Lp1a from the theraphosid *Lasiodora parahybana*). The number of lysines and arginines per cysteine on family level ranges usually between 0.2 and 1.1 corresponding to 4–16% of all amino acids (Table 9). Mini-proteins with a low content of lysine and arginine (< 8% of all amino acids) can be found in Hexathelidae, Agelenidae, Amaurobiidae and Sparassidae. Much higher contents of lysine and arginine occur in Theraphosidae and Ctenidae (11% as family average), 13% in Lycosidae and 16% in *Cupiennius salei* (Table 9).

Low numbers of lysine and arginine are usually sporadically positioned at the C- and N-terminal side, whereas high numbers are mostly concentrated at the C-terminal side. The last 10 amino acids of CSTX-1 (ω-ctenitoxin-Cs1a) from *Cupiennius salei* are KGKKKNKKTK-NH_2, of U_7-ctenitoxin-Pr1a from *Phoneutria reidyi* are EKKKKECPNV-NH_2 (Richardson *et al.*, 2006) and of U_{32}-theraphotoxin-Cj1a from the theraphosid *Chilobrachys guangxiensis* are KKAKCKHKSG-NH_2 (Chen *et al.*, 2008a), thus these four to seven lysines represent a charged C-terminal side. The lysine rich C-terminal part of CSTX-1 is responsible for its toxicity (Kuhn-Nentwig *et al.*, 2000). Less commonly, the

TABLE 9

Characteristics of mini-proteins with ICK motif for major spider taxa

	Number of toxins recorded per given number of cysteines									
	6	7	8	9	10	11	12	14	Number of Lys and Arg per Cys	% Lys and Arg of total number of amino acids
Hexathelidae	61	1	18		8				0.33 ± 0.24	6.2 ± 5.4
Theraphosidae	122	4	27		5	1			0.6 ± 0.29	11.0 ± 5.5
Other mygalomorphs	2	1	8						0.46 ± 0.12	7.8 ± 4.0
Haplogynes		1	14		12		7		0.45 ± 0.29	9.2 ± 4.1
Agelenidae	28	2	25		2		7		0.33 ± 0.23	7.4 ± 4.4
Amaurobiidae			4						0.16 ± 0.16	4.4 ± 4.5
Sparassidae	7								0.33 ± 0.14	7.7 ± 3.3
Ctenidae	14	7	10	7	13	2	8		0.47 ± 0.18	11.5 ± 4.3
Cupiennius			15		1				1.05 ± 0.46	16.3 ± 4.7
Lycosidae		9	111	8	18	1	5	6	0.81 ± 0.29	13.1 ± 3.3
Oxyopidae					3				0.8 ± 0.17	14.7 ± 1.1
% of peptides	38.2	4.1	39.2	2.4	10.1	0.7	4.4	1.0		

Ctenidae excludes *Cupiennius*.

lysine rich part may also be found in more central parts of the peptide as the positions 12–25 with seven lysines of U_1-theraphotoxin-Hh1a from the theraphosid *Haplopelma schmidti* show (KEGDKPC KKKKCK; Jiang *et al.*, 2008). In rare cases, lysine and arginine are more or less regularly distributed over the peptide, as lasiotoxin-1 = U_1-theraphotoxin-Lp1b from the theraphosid *Lasiodora parahybana* shows (FFECTFECDI KKEGKPCKPK GCKCKDKDNKDHKK CSGGWR CKLKLCLKF; Escoubas *et al.*, 1997). This distribution of charges leads to charged areas in the folded peptide that are important for the binding process of the target receptor. Very high amounts of charged areas as in the last case of *Lasiodora parahybana* raise the question whether this peptide is still neurotoxic or already in transition to a membrane active peptide.

With increasing number of cysteines and increasing structural complexity, the molecular mass of mini-proteins also increases, on average from 4 of a 6-cysteine peptide to 8 kDa for a 12-cysteine peptide. This corresponds to an average increase in molecular mass of 670 Da/additional cysteine. Nevertheless, several spider taxa follow different strategies. The average theraphosid and *Cupiennius salei* peptide size is always larger than that from hexathelid, agelenid or ctenid peptides, when referred to the same cysteine content. Lycosids again have a different strategy by generally keeping their mini-proteins small. They are more compact than the other peptides and their average increase in molecular mass is only 230 Da/additional cysteine (Fig. 10).

3.6 LARGE PROTEINS

The largest compounds in spider venom are proteins with a molecular mass between 110 and 140 kDa. They have so far exclusively been found in the genera *Latrodectus* (black widows), *Steatoda* (a species-rich genus where some

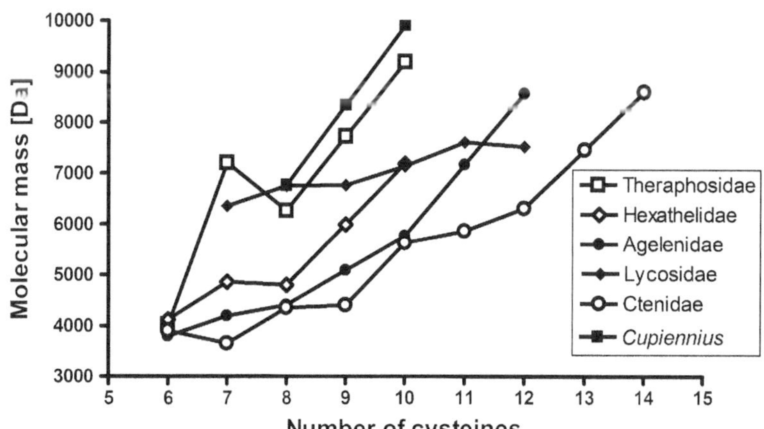

FIG. 10 Molecular mass of neurotoxic peptides depending on the number of cysteines for six spider families (Ctenidae without *Cupiennius*) and *Cupiennius salei*.

species superficially resemble black widows, which is why they are called false black widows) and *Achaearanea* (many species, among them the common house spider *Achaearanea tepidariorum*; Mironov *et al.*, 1986). These genera belong to Theridiidae, also known as the tangle-web spiders, cobweb spiders, or comb-footed spiders, and they represent the fifth-species richest family of spider species. Theridiids build specialised webs containing sticky droplets and they achieved many further specialisations with respect to prey, web or social life within their evolution. They are closer related to araneid and nephilid orb weavers (Fig. 2).

Investigations into the venom of black widow spiders (*Latrodectus*) concern primarily three species, the North American L. *mactans*, the Eurasian *L. tredecimguttatus* and the Australian L. *hasselti*. In all cases, the venom consists of seven major proteins selectively toxic to three groups: (1) α-latrotoxin (α-LTX) that is vertebrate selective and has a molecular mass of 130 kDa. (2) Five latroinsectotoxins (LITs), named α-, β-, γ-, δ- and ε-LIT, which are selective for insects, and have a molecular mass between 110 and 140 kDa. (3) The 120-kDa α-latrocrustotoxin (α-LTX) that is selective for crustaceans (Grishin, 1998). These masses correspond to 1000–1200 amino acid residues with a rather high level of homology of over 30% residue identity (Vassilevski *et al.*, 2009).

α-LTX contains three distinct domains and has a high affinity to form dimers that aggregate into tetramers and then insert into a lipid membrane of nerve cells, thus forming a central channel. This pore acts as a non-selective cation channels allowing a massive influx of extracellular Ca^{2+} into the nerve, which leads to vesicular exocytosis (Rohou *et al.*, 2007). In other words, this represents an exhaustive neurotransmitter release from a variety of nerves that depletes the synaptic vesicles, blocks the signal transmission, and causes muscular paralysis. α-LTX causes secretion of all known neurotransmitter types and the effects of all other toxins from *Latrodectus* (LITs and latrocrustotoxin) are rather similar. The highly toxic effect of the *Latrodectus* toxins is enhanced by a 8-kDa peptide that is not toxic and cannot form membrane pores by itself, but augments the affinity of α-LTX to the plasma membrane (Gasparini *et al.*, 1994; Kiyatkin *et al.*, 1995).

Rohou *et al.* (2007) also mentions the difficulties to distinguish the different toxic proteins within one *Latrodectus* species and between separate species. Indeed, the principal components of different venoms are very similar, sometimes even impossible to discriminate by SDS-PAGE and immunological assays. The same holds also for proteins from the venoms of *Steatoda paykulliana* and *Steatoda grossa* (Cavalieri *et al.*, 1987; Graudins *et al.*, 2002a) that are very similar to LITs and latrotoxin, respectively. In addition, Isbister and Gray (2003) described the clinical symptoms after bites of professionally identified *Latrodectus* and *Steatoda* spiders as indistinguishable and those of *Achaearanea* bites as similar. This may lead to the conclusion that all three genera, which are closely related within Theridiidae, rely in a similar way on neurotoxic proteins.

3.7 ENZYMES

3.7.1 *General aspects*

There have been long discussions as to whether spider venom can contain enzymes (Rash and Hodgson, 2002). In the past, venom was mainly taken either from spider prosoma extracts, from dissected glands or from gland punctures, bearing the danger of contamination with tissue proteases. Alternatively, venom obtained by electric milking may result in contamination with digestive fluid if not carefully performed. This has been broadly discussed by Perret (1977) who came to the conclusion that proteolytic activity in spider venom generally is due to contamination. Kuhn-Nentwig *et al.* (1994), however, showed that carefully cleaning of the chelicerae prior to electric milking is appropriate to avoid digestive fluid contamination for *Cupiennius salei*. If enzymes are still detectable in spider venoms even after such diligent procedure, then they are a constitutive component of spider venom. One must, however, remember that electric stimulation may also affect the enzymatic activity.

This view is supported by 50-years-old results indicating the presence of specific enzymes such as hyaluronidase and collagenase in spider venoms (Bettini, 1978; Kaiser, 1956; Schanbacher *et al.*, 1973b). Today, enzymes from spider venoms are known from 14 families with 49 species (Table 10). Most of these enzymes can be separated into two groups, one that cleaves polymers in the extracellular matrix, and the other one targeting phospholipids and related compounds in membranes. The overall purpose of such an enzyme arsenal co-injected with toxins into a prey's tissue looks obvious: by destroying the barrier of extracellular matrix and cell membranes the toxins can faster reach their targets. Additionally, the proteolytic activity of some of these enzymes facilitates the subsequent preoral digestion.

A few other enzymes have been found in spider venoms. A peptide isomerase from the agelenid *Agelenopsis aperta* serves to alter the chirality of Ser46 to the D-configuration in ω-agatoxin IVB (ω-agatoxin-Aa4b), which improves the activity of the toxin by nearly two orders of magnitude (Heck *et al.*, 1994; Shikata *et al.*, 1995). An ATPase catalyzing the dephosphorylation of ATP to ADP has been described from the venoms of *Loxosceles laeta* and *L. reclusa* (Geren *et al.*, 1976; Schenone and Suarez, 1978). Kininase, a bradykinin-inactivating enzyme, has been isolated from the venom of *Latrodectus tredecimguttatus*. It could be identified as a thiol-endopeptidase cleaving with high specificity the Pro-Phe bond of bradykinin and angiotensin I (Akhunov *et al.*, 1996). Acid and alkaline phosphatases are involved in a variety of biological processes, as signal transduction, metabolic regulation and energy metabolism (Kostrewa *et al.*, 1999). They were identified in the araneid *Parawixia bistriata* (Rodrigues *et al.*, 2006) and in the sicariid *Loxosceles reclusa* (Heitz and Norment, 1974).

TABLE 10
Enzymes known from spider venoms

Family	Species	Enzyme
Agelenidae	*Agelenopsis aperta*	Peptide isomerase
Araneidae	*Eriophora transmarina*	Collagenase (proteinase)
	Parawixia bistriata	Acid phosphatase, alkaline phosphatase
Ctenidae	*Cupiennius salei*	Hyaluronidase
	Phoneutria, two species	Hyaluronidase, protease
Desidae	*Badumna insignis*	Hyaluronidase, protease
Eresidae	*Eresus kollari*	Phospholipase A2
Hexathelidae	*Atrax robustus*	Hyaluronidase, phosphodiesterase, protease
	Hadronyche versuta	Phospholipase A
Lamponidae	*Lampona cylindrata*	Hyaluronidase, protease
Lycosidae	*Hippasa*, three species	Hyaluronidase, metalloprotease, serine protease, phospholipase A2
	Lycosa, two species	Hyaluronidase, protease (collagenase) serine protease
Miturgidae	*Cheiracanthium mildei*	Phospholipase A2
Nephilidae	*Nephila edulis*	Collagenase (peptidase)
Sicariidae	*Loxosceles*, 16 species	Alkaline phosphatase, ATPase, esterase, hyaluronidase, metalloproteinase, protease, serin-protease, phospholipase D, chitinase
	Sicarius, 9 species	Phospholipase D
Sparassidae	*Holconia immanis*	Collagenase (proteinase)
Theraphosidae	*Aphonopelma*, three species	Hyaluronidase, phosphodiesterase, protease
	Vitalius dubius	Hyaluronidase
Theridiidae	*Latrodectus tredecimguttatus*	Hyaluronidase, kininase, phosphodiesterase

Spider families in alphabetic order, enzymes are mainly mentioned as described in the literature, not correcting for the fact that some of the terms refer to higher or lower classification levels.

3.7.2 *Enzymes targeting the extracellular matrix*

Hyaluronidases, historically the first class of enzymes found in spider venoms, cleave the mucopolysaccharide hyaluronic acid, a major constituent of the extracellular matrix (Kaiser, 1956). It sounds obvious that, by cleaving this polymer, hyaluronidases facilitate the spread of toxins and further venomous compounds when co-injected into a prey's tissue; therefore it had been frequently termed ''spreading factor'' (e.g. Bettini, 1978). This enzyme may be slightly different between different spider species since its apparent molecular mass varies between 32 and 44 kDa (Barbaro *et al.*, 2005; da Silveira *et al.*, 2007a; Nagaraju *et al.*, 2007a; Young and Pincus, 2001). Hyaluronidases have been reported from the venom of 20 spider species from eight families: Hexathelidae (Sutherland, 1978), Theraphosidae (Rocha-e-Silva *et al.*, 2009; Savel-Niemann, 1989; Schanbacher *et al.*, 1973a), Sicariidae (Barbaro *et al.*, 2005; da Silveira *et al.*, 2007a; Geren *et al.*, 1976; Schenone and Suarez, 1978; Wright *et al.*, 1973; Young and Pincus, 2001), Desidae (Schenone *et al.*, 1989), Lamponidae (Rash and Hodgson, 2002; Young and Pincus, 2001), Theridiidae (Bettini and Maroli, 1978), Ctenidae (Kaiser, 1956; Kuhn-Nentwig *et al.*, 1994), Lycosidae (Kaiser, 1956; Nagaraju *et al.*, 2006, 2007a; Rash and Hodgson, 2002) and they are probably wider spread. According to Schanbacher *et al.* (1973a), hyaluronidase is one of the major constituents of the venom of the theraphosid *Aphonopelma hentzi*.

Collagenases are matrix metalloproteases cleaving peptide bonds between proline and other amino acid residues in collagen, a key compound in the animal extracellular matrix. These and other metalloproteases have been detected in the venom of Araneidae, Nephilidae, Sparassidae (Atkinson and Wright, 1992) and Lycosidae (Nagaraju *et al.*, 2007b). Also, the venoms of different *Loxosceles* species have been well investigated with respect to their proteolytic activities (Barbaro *et al.*, 2005; da Silveira *et al.*, 2002; Feitosa *et al.*, 1998). *Loxosceles* astacin-like proteinases (LALP, 32–35 kDa) have been identified in the venom as well as by transcriptome analysis and more than three isoforms have been detected in *Loxosceles intermedia*. Remarkable is the homology with digestive astacin-like metalloproteinases from the crawfish *Astacus astacus*, which could be interpreted that venomous LALPs first act as spreading factor but then facilitate the subsequent degradation of the prey by preoral digestion (da Silveira *et al.*, 2007b; Gremski *et al.*, 2010; Trevisan-Silva *et al.*, 2010). From the same species, also two large serine proteases (85 and 95 kDa) were detected and they may be important in local tissue destruction (Gremski *et al.*, 2010; Veiga *et al.*, 2000). Interestingly, two additional serine proteases (16.3 and 28.7 kDa) from the lycosid *Hippasa agelenoides* (Devaraja *et al.*, 2008, 2010a,b) have been reported to interfere in hemostasis. Transcriptome analysis of the venom glands of *Lycosa singoriensis* reveals a cDNA sequence similarity to a serine protease cDNA (LIS237) from *Loxosceles intermedia* (Gremski *et al.*, 2010; Zhang *et al.*, 2010). Recently, in the venom gland transcriptome analysis of *Loxosceles laeta*, a cDNA encoding chitinase has been reported (Fernandes-Pedrosa *et al.*, 2008).

A variety of proteases have been detected in spider venoms. Depending on the kind of test applied, the determination of these enzymatic activities is more or less advanced. This gives a picture according to which spiders may apply unspecialised proteinases as well as rather specific ones. Further proteases with molecular masses between 18 and 33 kDa were described from venoms of Hexathelidae, Theraphosidae, Lycosidae, Desidae and Lamponidae (Rash and Hodgson, 2002).

3.7.3 *Enzymes targeting membranes*

Cell membranes of living organisms consist of a lipid bilayer that is mainly composed of phospholipids. Inner and outer surfaces differ in chemical composition with phosphatidylcholine, sphingomyelin and a variety of glycolipids determining the outer surface. These molecules are the targets of various hydrolase enzymes breaking down the phosphodiester bond (phosphodiesterase), degrade sphingomyelin (sphingomyelin phosphodiesterase or sphingomyelinase D), or hydrolyses phospholipids (phospholipases).

Up to now, protein toxins from sicariid spiders that are involved in dermonecrosis and massive inflammatory response, have been named dermonecrotic toxins and are well reviewed by Rash and Hodgson (2002). Cunha *et al.* (2003) named such a dermonecrotic toxin loxnecrogin but in general these proteins are better known under their enzymatic activity description as sphingomyelinase D (Kurpiewski *et al.*, 1981; Tambourgi *et al.*, 1998). Lee and Lynch (2005), followed by Chaves-Moreira *et al.* (2009), propose to replace the term sphingomyelinase D by PLD because this enzyme also hydrolyzes lysoglycerophospholipids or lysophosphatic acid. PLDs are 31–34 kDa enzymes, identified so far in the venom of 16 *Loxosceles* species and 9 *Sicarius* species (Barbaro *et al.*, 2005; Binford and Wells, 2003; Binford *et al.*, 2008). Remarkably, PLDs from different *Loxosceles* species are very similar in terms of molecular masses, amino acid sequences and biological activities; and several isoforms have been described (Binford *et al.*, 2009; Gremski *et al.*, 2010; Kalapothakis *et al.*, 2007).

Within sicariids spiders, several other enzymes and some low molecular mass compounds have been detected, but only four insecticidal peptides of unknown target with an ICK motif in *Loxosceles intermedia* (de Castro *et al.*, 2004; see also Section 3.5.2.1). Interestingly, one cDNA of *Loxosceles intermedia* venom glands shows similarity to the neurotoxin Magi-3 (μ-hexatoxin-Mg2a) from the hexathelid *Macrothele gigas* exhibiting specific inhibitory effects towards the insect sodium channel (Corzo *et al.*, 2003; Gremski *et al.*, 2010). Nevertheless it is possible that sicariids are the spider family relying to the highest degree on the activity of enzymes when subduing a prey (Binford *et al.*, 2009).

Since PLDs were only known from microorganisms and sicariids spiders, but not from other animals, Cordes and Binford (2006) speculated about horizontal gene transfer from *Corynebacterium* species to ancient sicariid spiders. However, PLDs have meanwhile also been identified in ticks and a convergent evolution in distinct lines of this class of enzymes is postulated rather than a horizontal gene transfer (Fry *et al.*, 2009). Other phospholipases targeting the phospholipid compounds in membranes have been found in four families: Hexathelidae (Sheumack *et al.*, 1984), Eresidae (Foradori *et al.*, 2005; Usmanov and Nuritova, 1994), Lycosidae (Nagaraju *et al.*, 2006), Miturgidae (Foradori *et al.*, 2005), but it will certainly have a much wider abundance among spiders. Additionally, phosphodiesterase activity was identified in *Atrax robustus*, *Aphonopelma cratus*, *Latrodectus mactans* (Russell, 1966) and esterase activity in *Loxosceles reclusa* (Wright *et al.*, 1973).

In humans, necrotic cutaneous lesions or skin ulcerations are known since long time but only since 60 years it is confirmed that one of the possible reasons are spider bites, especially from spiders of the genus *Loxosceles* and (rarely) *Sicarius*. Today, PLDs, up to now unique to sicariids, are seen as main causative for this degenerative process, and "necrotic arachnidism" (sensu Rash and Hodgson, 2002) is therefore more or less restricted to sicariid bites. More in general, however, spiders injecting large quantities of enzymes and thus destroying tissue may also cause necrosis or ulceration. Especially collagenases have been mentioned as further enzyme causing necrosis (Atkinson and Wright, 1992), but a comparison of the necrotic potential of 45 spider venoms from 20 families yielded only necrotic evidence for *Loxosceles* venom and showed that this phenomenon has widely been overestimated (Foradori *et al.*, 2005).

4 Discussion

4.1 VENOM EVOLUTIONARY STRATEGIES

Spiders possess venom glands to produce venom that they use to paralyse and/or kill their prey items or as a weapon against predators. There seems to be one main strategy to reach these goals and this relies mainly on mini-proteins. Most spiders investigated so far possess a variety of mini-proteins in their venom glands, and we have to assume that this represents the plesiomorphic repertoire of all spiders, that is the first, most ancient equipment. In contrast to this assumption, it could also be argued that other venom components (e.g. the lower molecular mass compounds or enzymes) represent the plesiomorphic situation. However, as long as we have no information on the venom composition of the most plesiomorphic spider group (Mesothelae) and no contradicting information from the venoms of mygalomorph families, we stick to the most simple and most convincing assumption that the today most widely spread venom mechanism also was the most plesiomorphic one.

This plesiomorphic repertoire of venom components consists of several mini-proteins acting primarily on ion-sensitive channels and receptor proteins. It seems to be well functioning and reliable, but nevertheless was permanently modified during the evolution of spiders. Besides mini-proteins further groups of compounds have been added to the venom machinery to support its overall function, several types of enhancers have been developed and a few spider families apparently even replaced mini-proteins completely with unrelated substances having a different mode of action. Thus, during spider evolution, the initial venom composition was permanently modified and it seems that the number of compounds involved increased.

4.1.1 *Modifying mini-proteins*

Structurally mini-proteins can be considered as superficially similar peptides in the range of 3–8 kDa. However, the variation of only a few features such as molecular target (functional diversity), sequence variations, number of cysteines or size (structural diversity) yields a nearly endless number of different toxins. In many spider species, dozens of bioactives occur and the record so far consists of 166 mini-proteins known from one peptidomic study on *Haplopelma hainanum* (Tang *et al*., 2010). Quite often, the difference between the structures of two peptides consists only in an exchange of one amino acid, which can result in drastic changes of potency and/or selectivity. Nevertheless, among the nearly 1000 mini-proteins compiled for this study, not a single sequence was identified in two distinct spider species. Some spider families seem rather uniform with respect to size and structure of their mini-proteins (e. g. measured as number of cysteines: 6 in Sparassidae, 8 in Amaurobiidae, 10 in Oxyopidae), while other show a pronounced focus on lower or higher cysteine numbers or cover nearly the whole range (e.g. Lycosidae with 7–14 cysteines).

The reason for such an overwhelming variation enthusiasm can probably be seen in the co-evolution of spiders and other arthropods, which obviously always have been their main prey. This development of both groups lasts since the early Carboniferous period, since nearly 300 million years (Penney and Selden, 2011). Obviously spiders search permanently new target receptors for their toxins, experience combinations of the same venom compounds for different prey taxa or go for higher binding efficiency. The permanent change in the composition of prey taxa due to evolution and/or habitat change enforces subsequent changes in number, structure and function of venom bioactives.

We failed in detecting a strong phylogenetic context between modifications of mini-proteins and spider evolution. The described taxonomic context, for example Sparassidae having only six-cysteine mini-proteins, has to be considered very carefully, because research is highly biased focusing on a few spider families, primarily mygalomorphs. Some of the taxonomic correlates described may become weak in the future by the increasing number of mini-proteins known from more and differing species, some correlates may also be reinforced and confirmed, or even new ones may show up.

4.1.2 *Assisting mini-proteins*

A variety of different compounds, namely low molecular mass compounds, linear peptides (α-SCP) and enzymes frequently co-occur with mini-proteins. They address two different modes of supporting neuroactive peptides (1) by directly attacking neuronal or muscular cells or (2) by destroying target tissue so that the mini-proteins have a higher probability to reach their target.

(1) Between low molecular mass cationic compounds, SCP, many neurotransmitters and neuromodulators are among the most remarkable compounds. Virtually any possible type of substances known to interact with neurons has been identified in spider venoms. The spectrum reaches from neuroactive amino acids, biogenic amines and polyamines to acetylcholine and noradrenaline. Since acetylcholine is the most abundant excitatory neurotransmitter of chemo- and mechanosensory

receptor neurons in the insect nervous system, it is obvious that the activity of these low molecular mass compounds directly adds to the neurotoxic effect of mini-proteins.

(2) α-SCP, α-LCP and enzymes have in common that they destroy cells and tissue of the target prey or victim. Both groups comprise high degrees of structural and functional diversities: α-helical peptides vary in length, cationic charge and function; enzymes target the extracellular matrix (among others, hyaluronidase and collagenase) or membranes (among others, phosholipase D). Historically, hyaluronidases have been named ''spreading factor'' but this term could be used for most linear cytolytic peptides and enzymes since their effects are similar: tissues of the prey item are destroyed and the other toxins can reach their targets faster and more efficiently.

From this, it can easily be concluded that energetic reasons may be a main driving force behind the inclusion of low molecular mass compounds, linear cytolytic peptides and enzymes into spider venoms. Even if we see in laboratory experiments, how effective and specialised a given mini-protein acts towards a given receptor or cell type, it makes a big difference to achieve the same result in a living animal, since several layers of different tissues usually prevents fast venom diffusion.

Here again, we failed to detect an obvious phylogenetic context within low molecular mass compounds or between these and other compounds in spider venom. Neurotransmitters are remarkably often encountered in Sparassidae (15 records) and in Theraphosidae (26 records), but especially in Theraphosidae, this is again certainly due to research bias. There is no apparent co-occurrence of membranolytical (cytolytical) substances and citric acid or of citric acid and Ca^{2+}-dependent enzymes (i.e. phospholipase). With a few exceptions, research in the field of low molecular mass compounds of spider venoms was for long time more a side-activity than in the focus of most research groups. Furthermore, experts in peptides and proteins tend not to have the required expertise and/or appropriate analytical instrumentation to investigate such molecules, not made of simple amino acids. One of the astonishing outcomes of this review is the high diversity of low molecular mass compounds in spider venom, being the second most frequent compound group (after mini-proteins) and comprising nearly one sixth of all records. We feel that many more compounds can be detected and we would not be surprised if phylogenetic and functional constraints could be detected. It is also astonishing to realise the broad spectrum of enzymes detected in spider venoms and the frequency with which these can be recorded from all spider families. With the remarkable exception of Sicariidae (see below); however, we could not detect any relevant phylogenetic context.

Indeed, such a remarkable phylogenetic context exists for linear peptides (α-SCP, α-LCP). These are common in Zodariidae and Lycosoidea (Ctenidae,

Oxyopidae, Miturgidae and Lycosidae) and obviously very successful. In *Cupiennius salei* they are even more common than mini-proteins. α-SCPs and α-LCPs have so far not been detected outside the two mentioned spider taxa and must therefore be regarded as their new achievement. Additional records of the cytolytic activity of crude venom from other spider families, tested on various cultured cells (Cohen and Quistad, 1998), are probably due to side effects of other venom components, but again stress the point of more targeted research into this class of compounds. If some of these additional records should be proven (compare Table 5), linear cytolytic peptides could be seen as new achievement of entelegyne spiders or at least parts of them.

4.1.3 *Synergizing mini-proteins*

An effect is enhanced or synergistic when it is caused by the combination of two or more compounds that create a resulting effect greater than the sum of both separately. Among toxin compounds, usually a combination of one mini-protein and low molecular mass compounds or a combination of two mini-proteins, can act synergistically. For a variety of spider venoms, such synergistic combinations are known that increase the toxicity of the mini-protein considerably. This enables the spider to inject a lower quantity of venom to achieve the same result. The drivers behind this process are energy saving reasons.

In *Cupiennius salei*, the toxic function of the main toxins CSTX-1 (ω-ctenitoxin-Cs1a) and CSTX-9 (U_1-ctenitoxin-Cs1a) is enhanced by CSTX-13 (U_2-ctenitoxin-Cs1a), which itself has only a negligible toxicity (Wullschleger *et al.*, 2004, 2005). All three peptides share the common ICK motif and CSTX-13 probably developed from an ICK-toxin to an ICK-enhancer without own toxicity. Also a variety of low mass compounds enhance the toxicity of CSTX-1, namely the high K^+ ion content of the venom, histamine and the cytolytic peptide cupiennin 1a. Altogether, this saves up to 50% of the main toxic compound (Wullschleger *et al.*, 2005). Alternatively, CSTX-13 may further have another, yet unidentified, bioactivity.

The activity of α-LTX from *Latrodectus tredecimguttatus*, a huge 130 kDa protein, also seems to depend on latrodectin (Grasso and Pescatori, 1996), a 8-kDa peptide that is not toxic on its own and cannot form membrane pores, but stabilizes the structure resulting in a higher biological activity of α-LTX (Gasparini *et al.*, 1994; Kiyatkin *et al.*, 1995).

4.1.4 *Replacing mini-proteins*

If mini-proteins can be substituted by something better, it is obvious that they should be replaced. ''Better'' may refer to less energetic input or improved efficacy. There are at least four examples of spider families that more or less completely replaced mini-proteins by ''better'' compounds.

(1) Amino acids containing acylpolyamines are only found in Araneidae and in Nephilidae, two families of orb-weaving spiders (Fig. 2). Since there is no obligatory co-occurrence of acylpolyamines with potential enhancers such as neurotransmitters, it must be concluded that the venomous effect in araneids and nephilids is primarily due to these acylpolyamines. Because of the positive charges on the amines of their polyamine chain, these compounds block cationic-selective ion channels, mainly Ca^{2+} channels and K^{+} channels (Mellor and Usherwood, 2004), thus they can be seen as an interesting equivalent to mini-proteins.

In Araneidae and Nephilidae, at least two further features support the efficiency of the venom. These spiders have a very sophisticated prey wrapping behaviour prior to biting (Araneidae) or are generally large and powerful enough to directly bite their victims (Nephilidae; Robinson and Mirick, 1971; Robinson *et al.*, 1969). Secondly, after having completed their orb web, species from both families apply droplets with a sticky substance to the spiral thread in the orb web. This substance derives from the aggregate glands, a modified silk gland. The aqueous solution of the sticky substance contains a variety of precursors of neurotransmitters such as choline, *N*-acetyl-taurine and GABA (Vollrath *et al.*, 1990), also several saturated and unsaturated fatty acids, including polyunsaturated fatty acids were found. Astonishingly, rare neurotoxic compounds such as tetrahydro-β-carboline and an organometallic 1-(diazenylaryl)ethanol could also be identified (Marques *et al.*, 2004, 2005; Salles *et al.*, 2006). It is obvious that the sticky substance of orb weavers not only traps prey items but also paralyzes them. These two families of orb-weaving spiders replaced mini-proteins by acylpolyamines, but simultaneously enhanced the power of this new venom by behavioural adaptations and toxic compounds in their sticky glue. This combination seems to compensate for the fact that araneid and nephilid venoms paralyze their victims very efficiently but do not kill them (Nentwig *et al.*, 1992; Salles *et al.*, 2006).

From a phylogenetic point of view, it is highly interesting to know how the venom of spiders from other orb-weaving families and close relatives is organized. Tetragnathidae is another spider family with orb-webs containing a sticky substance and closely related to Nephilidae and Araneidae, but nothing is known about their venoms or the composition of their sticky droplets. Theridiidae and Linyphiidae are two very species-rich families with different web types, which both may also contain a sticky substance. Nothing is known on its chemical composition and the venom of Linyphiidae has never been investigated.

Besides the here mentioned amino acids containing acylpolyamines a variety of spider venoms include non-amino acids containing acylpolyamines, namely Agelenidae, Amaurobiidae, Ctenizidae, Hexathelidae, Pisauridae and Theraphosidae. With the exception of Ctenizidae and Pisauridae, where the venoms are hardly investigated, in all other taxa mini-proteins obviously constitute the main compounds of the venom. Thus, it seems that non amino acids containing acylpolyamines obviously play a less central role than the amino acids containing acylpolyamines in the venoms of Araneidae and Nephilidae.

(2) Though Theridiidae belong to the same superfamily Araneoidea as orb-weaving spiders, their venom is completely different and they replaced mini-proteins by large neurotoxic proteins. Unfortunately it is not known if this replacement is true for all theridiid spiders since only three *Latrodectus* species have been analysed. For further three species from two related genera there is evidence for similar venom composition, the >2300 other theridiid species remain uninvestigated so far.

(3) All sicariid spiders, that is *Loxosceles* and *Sicarius* species, contain phospholipases D in their venom, an enzymes unique to spiders (Cordes and Binford, 2006). These enzymes are very potent in degrading sphingomyelin, thus destroying cell membranes. Therefore, it had been assumed that sicariids replaced mini-proteins by PLDs. Meanwhile, six insecticidal peptides with an ICK motif have been identified from *Loxosceles intermedia* venom (de Castro *et al.*, 2004), again indicating a research gap due to narrowed research focusing on PLDs. Nevertheless, it is still probable that sicariids are the spider family relying to the highest degree on enzymatic activity when subduing prey (Binford *et al.*, 2009).

(4) Two-domain modular toxins occur in three families and each case is unique. These recent findings could be regarded as exceptions, but we assume that more examples will be detected soon. The venom of *Cheiracanthium* (Miturgidae) contains a large peptide forming amphipathic structure effectively interacting with membranes and causing a pronounced cytolytic effect. This peptide obviously derived from a typical neuroactive peptide since it contains two distinct primary structures following the ICK motif, but it does not seem to interact with ion channels and the overall sequence homology with other ICK-peptides is very low (Vassilevski *et al.*, 2010). In *Lachesana* (Zodariidae), the venom consists of several α-SCPs and some α-LCPs composed of two short peptides in a "head-to-tail" orientation (Vassilevski *et al.*, 2008). Astonishingly, in both taxa the duplication of the initial structure, whether a neuroactive peptide or small cationic peptide, was the successful step to achieve a different and obviously very efficient mode of action. In contrast to this, the "double-knot" toxin DkTx of the theraphosid *Haplopelma schmidti* still maintained its neurotoxic function (Bohlen *et al.*, 2010). This indicates that "molecule-doubling" is no specific feature of higher evolved spiders.

4.2 WHICH VENOM STRATEGY IS MOST SUCCESSFUL?

4.2.1 *Evolutionary strategies*

When comparing the composition of different venoms, for example on the basis of HPLC-chromatograms produced under comparable conditions (Fig. 11), we realized that (1) all venoms are very different from each other, (2) all spiders have complex venoms (many chromatographic peaks), (3) some spider venoms are more or less rich in compounds (e.g. *Viridasius* vs. *Cupiennius*) and (4) even within the same family venoms can be very different (e.g. *Polybetes* vs. *Holconia*). Extensive intra-species or even intra-specimen variation studies would certainly reveal further complexity. From an evolutionary point of view, we have to assume that all spiders are as well adapted to their respective needs as possible, and that therefore all different venom strategies involved are successful.

Spiders have to run a successful venom strategy in all situations and we assume that the venom strategy of a mygalomorph spider species is as successful as the strategy of a haplogyne or entelegyne araneomorph spider. However,

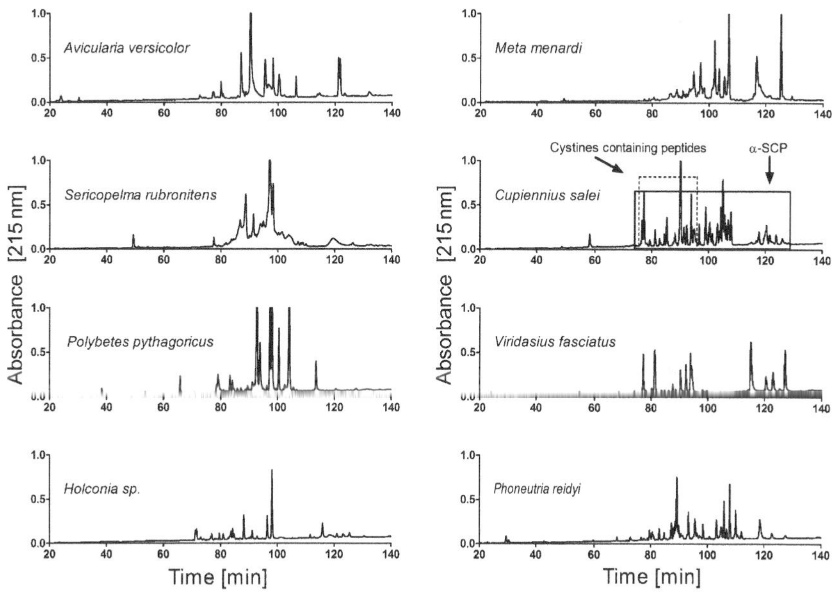

FIG. 11 RP-HPLC-chromatograms produced under comparable conditions for venoms of the theraphosids *Avicularia versicolor* and *Sericopelma rubronitens*, the sparassids *Polybetes pythagoricus* and *Holconia* sp., the tetragnathid *Meta menardi*, and the ctenids *Cupiennius salei*, *Viridasius fasciatus* and *Phoneutria reidyi*. The separation was done on a NUCLEOSIL 100-5 C_{18} column (Macherey & Nagel, 250/4 mm) using solvent A (0.1% trifluoroacetic acid in double distilled water) from 0 to 10 min, followed from 10 to 150 min by a gradient (0–60%) of solvent B (0.1% trifluoroacetic acid in acetonitrile).

As measured in an ELISA using monoclonal antibodies directed against venom compounds, *Cupiennius salei* injects only low amounts of venom into sensitive dipterans and crickets, whereas less sensitive carabid beetles receive a much higher venom quantity (Wigger *et al.*, 2002). With a different approach, also with *Cupiennius salei*, Malli *et al.* (1999) showed that the amount of venom injected into crickets depends on the size of crickets and on their activity: large items receive more venom than small items and crickets that are still active get more venom than motionless items. Since spiders in general are able to control the amount of venom injected, they usually first inject a small quantity of venom, wait a while and if the target is still struggling, add another dosis. This may be repeated several times and prevents to invest too much venom. *Cupiennius salei* knows very well the remaining amount of venom it its glands and by chemical cues also has a rather good idea on the identity and venom demand of a given prey target, both presenting important information for a decision on whether it shall attack or not a given target (Hostettler and Nentwig, 2006). This framework enables *Cupiennius salei* to optimize its venom investment into a target, indicating that venom is a precious resource, extremely important for survival and not regenerating in the short term (Wigger *et al.*, 2002). Though these findings derived only from one species, we believe that the principle can be generalized for most or even all spiders.

4.3 How many toxins are in spider venoms and why?

One of the most astonishing characteristics of spider venom is the high diversity of its compounds. Naively, one would expect that one toxic substance or a very few compounds would be sufficient to guarantee a high efficiency of the venom. In sharp contrast to this and as for most venomous animals, spider venom consists usually of several major groups of compounds and in total contains from several dozens to several hundreds of bioactive compounds.

Today, mainly peptidomic, transcriptomic and genomic investigations give information on up to several hundred peptides in the venom from one spider species. Liang (2004) mentions about 400 peptides and proteins from the venom of the theraphosid *Haplopelma schmidti*, more than 300 of them not characterized and he calls it ''tip of the iceberg''. Other legendary examples are *Haplopelma hainanum* with 166 and *Lycosa singoriensis* with 154 mini-proteins in our database (Tang *et al.*, 2010; Zhang *et al.*, 2010). These examples concern mini-proteins showing an extremely high potential of variation. Often two peptides differ only in the exchange of one amino acid, or by a stepwise shortening of an amino acid tail several different peptides can be obtained. So far, no identical mini-protein has been found between two species, even if the degree of similarity may be very high. Therefore it has been assumed, that the number of peptides per species can be multiplied with the number of species to obtain the total amount of potentially existing toxins.

This consideration may hold for spiders that mainly rely on mini-proteins such as mygalomorph species and some other families, but it can probably not be accepted for all other taxa. In orb weavers (Araneidae and Nephilidae) mainly relying on acylpolyamines, one species shares several compounds with related species within the family, but not across family borders (Itagaki *et al.*, 1997; McCormick and Meinwald, 1993; Palma *et al.*, 1998) and this prohibits the simple multiplication procedure. It should also be mentioned that for most spider families only a few compounds are known and it is completely unknown whether they show the same high variation pattern as theraphosids and some other species. Linear peptides also show a high degree of variation and many different peptides are known from one species. However, there is hardly any information on the species-specificity of these compounds. In low molecular mass compounds, where much more compounds have to be assumed per spider venom, the degree of species-specificity is extremely low. For enzymes, the situation may be the opposite.

To account for this, we classify the spider families into three groups: one with a probably high number of compounds (all mygalomorph families, agelenids, lycosids, ctenids), one with a probably low number of venom compounds (Araneidae, Nephilidae, Theridiidae) and the rest with a probably medium number of compounds. By multiplying the 5000 species in the high compound category with an estimate of 200 compounds, by multiplying the more than 31,000 species in the medium category with 100 compounds per species and by multiplying the more than 5000 species in the low compound category with 50 compounds, we arrive at an estimate of 3.75 million compounds that could exist in the venoms of the currently known 42,055 spider species (species numbers according to Platnick, 2011) based on the current knowledge. This refers to an average of 90 compounds per species. We are aware of the very coarse method we applied, but we feel that such a procedure is much more appropriate that other estimates often simply stating ''tens of millions or more'' peptides. Interestingly, our estimates yields a result in the same range as similar estimates for scorpions (1500 species could have 100,000 peptides) or cone snails (500 species have an estimated 50,000 peptides; Zhu *et al.*, 2003) referring to 70 and 100 peptides per species.

It has often been argued that these numbers most likely remain underestimations, because the technological developments nowadays allows the identification of components present in tiny amounts and not easily detected in the background noise. For cone snails these numbers have nowadays clearly been revised towards larger numbers. Cone snails have some 800 species described, and the venom of *Conus consors* for example was recently found to contain more than 1700 biomolecules (Biass *et al.*, 2009; Dutertre *et al.*, 2010).

Why do spiders and scorpions or cone snails have such high toxin diversity? The continuous variation of approved peptide patterns and selection for new functions may yield peptides with an affinity to yet inaccessible receptors, thus enlarging the spectrum of target receptors, that is functional diversity.

If new peptides without yet identified target do not have negative aspects such as additional costs, they even can be maintained as a kind of insurance for future need or as material for future modification. This process increases the number of peptides permanently and has also aspects of preventing resistance of potential prey targets against a single venom compound. Therefore, spiders usually have a broad arsenal of venom compounds with different modes of action, for example mini-proteins and linear peptides (SCPs and α-SCPs) and enzymes as in *Cupiennius salei* (Kuhn-Nentwig *et al.*, 1994), or they have beside enzymes a variety of mini-proteins in one venom, interacting with several receptor types (e.g. the *Phoneutria nigriventer* venom contains several peptides affecting various types of Na^+, K^+ and Ca^{2+} channels; Borges *et al.*, 2009).

The high diversity of mini-proteins represents a combinatorial library of biologically active molecules and this can probably also be applied to linear peptides and other venom bioactives such as acylpolyamines, large proteins or enzymes. The principle of combinatorial chemistry fuels an optimization process as part of the evolutionary arms race between predator and prey. Especially with respect to the composition of the spider venom, the result of this can be understood as accelerated evolution.

4.4 VENOM AS DIGESTION SUPPORT?

Since it has been established that enzymes and α-SCPs/α-LCPs are regular and frequent components of spider venoms, it is controversially discussed up to what degree venom also supports digestion (Minton, 1974). The usual arguments in favour of the ''venom digestion theory'' comprise the ability of the most common enzymes in spider venom (such as hyaluronidases and several proteases) to destroy tissue. Also linear peptides (α-SCP and α-LCP), by targeting at membranes, destroy the prey's tissue. But is this process enabling other venom compounds to reach their specific targets (usually neurons and muscle cells) sufficient to play ''a central role'' in the external digestion as stated by Vassilevski *et al.* (2009)?

It is certainly wrong that spider venom glands derived from digestive secretion glands or tissues that after developing the production of toxins simply maintained the enzyme production as assumed by Minton (1974). While the digestive tract of spiders is of mesodermal origin, venom glands are ectodermal invaginations. Also, digestive fluid consists of a variety of rather unspecific enzymes, produced in high amounts, and the secretion cells for digestive enzymes regenerate completely within 24 h (Foelix, 2011) while many venom enzymes are more specific and the regeneration time of venom is 8–16 days (Boevé *et al.*, 1995). Moreover, spiders use their venom very economically and avoid any waste (see Section 4.2.2, Wigger *et al.*, 2002), while digestive fluid is used massively. This quantitative argument is probably the most important, since already Atkinson and Wright (1992) showed that the collagenase activity in the venom of three spider species is less than 1% of the enzymatic activity of the midgut.

Nevertheless, enzymes and α-SCPs/α-LCPs from spider venom contribute to the destruction of the barrier of extracellular matrix and cell membranes, thus, principally supporting subsequent preoral digestion. After sufficient envenomation, most spiders heavily chew their prey, a process that is performed by the mouthparts consisting of endites (gnathocoxae) of the palps and the basal segment of the chelicerae, all usually heavily chitinised for this purpose. During this mechanical process the victim is permanently mixed with digestive fluid and this leads to a complete destruction of the prey's body. Also venom compounds are destroyed by digestive fluid. Especially linear peptides (α-SCP and α-LCP) immediately degrade after contact with trypsin/chymotrypsin like proteases due to the high content and distribution of lysine in the peptides. In conclusion, digestion is certainly supported by some venom compounds, but it is probably only of marginal importance.

4.5 RESEARCH AND SYSTEMATICS

Research on spider venoms suffers in two regards from an inappropriate attitude towards spider systematics. First, venom research focussed so far only on 0.41% of all known spider species, belonging to less than one third of the known families. This bias was driven by a selection towards large and easily accessible spiders and species of medical importance, while other important and large groups were completely neglected. It does not really matter whether the motivation of venom investigation is of pharmaceutical nature, intends to find insecticidal substances, understand physiological processes or targets in systematics. We believe that these different fields of research are highly interconnected and that therefore a wider view on the spider diversity is more than necessary.

Second, the use of valid names is essential (see Table 2). It is important to adopt the valid spider name even if publications with older or wrong names are available because this is the only way to correctly pin results to valid names. Since many years, access to spider taxonomy is particularly easy and convenient because Platnick's catalogue (Platnick, 2011) provides the valid nomenclature. If a researcher owns an unidentified species it is of outstanding importance to identify this species. Identification only to genus level is not sufficient. More than 10% of all species for which we collected venom compound records refer only to genus level and this yields quite often an uncomfortable situation. Admittedly, identification of spider species is not trivial and in the majority of cases needs an expert. If it is not possible to receive proper species identification, at least a voucher specimen should be deposited in a museum. Whenever possible, pictures of the animals and biotope should be taken and the precise geographic origin should be documented. All this information should be mentioned in a subsequent publication where this species is involved.

4.6 CONCLUSIONS

In conclusion, our investigations clearly illustrate that spider venom components still have many original bioactivities to reveal, only a minor portion of the huge molecular biodiversity they offer has been studied. Venoms have mostly been studies when originating from species harmful to humans (*Phoneutria*, *Loxosceles* and *Latrodectus* species) or obtained from easily available, large and easy to milk specimen (Theraphosidae typically). Not only have a very few species been looked at (less than 1%), most of these were further only partially studied. Biological activities seem often to be found where researchers look at, or around what they are looking for. One should thus be careful with paradigms stating that a given venom has a specific activity or is composed "only" of one class of spider venom components. It may just be that one research team only, specialized in one aspect of toxinology, studied that particular venom. Are there rules or is this mostly the result of topic-focused research? A broader view is clearly needed and only future large scale studies will tell.

Our current knowledge suggests that spiders have evolved towards several objectives, but mainly as a result of two major considerations: The first is a permanent need to save energy and avoid any unnecessary investment. Spiders may not find food every day and its takes 1–2 weeks to regenerate the venom. As a result, the venom has gained in potency and synergistic mechanisms have been optimized. The second evolutionary pressure is correlated to the absolute need of efficacy when envenoming a prey or a predator. To this end, Nature has favoured increased complexity of venom to secure a bioactivity of a complex cocktail made of highly potent and selective ingredients that will affect a broad range of targets and offer the best adaptation to environmental changes.

Spider venoms represent what probably is one of the world's largest combinatorial compound library selectively developed to generate highly potent and selective bioactives that have undergone a natural lead optimization process through million years of natural selection. Venom composition and strategies in spiders: is everything is possible? We believe yes!

Acknowledgements

We thank Barbara Knoflach-Thaler, Christian Kropf and Mark A. Townley for discussing some topics from this manuscript with us and the Swiss National Science Foundation for funding. We are grateful to Dominique Koua (Atheris Laboratories, Geneva, Switzerland) for bioinformatics assistance. Some of the software used in this study was developed in the frame of "CONCO—The cone snail genome project for health", an integrated project funded by the European Commission, ref. LSHB-CT-2007-037592 (http://www.conco.eu).

References

Adams, M. E. (2004). Agatoxins: ion channel specific toxins from the American funnel web spider, *Agelenopsis aperta*. *Toxicon* **43**, 509–525.

Adams, M. E., Bindokas, V. P., Hasegawa, L. and Venema, V. J. (1990). ω-Agatoxins: novel calcium channel antagonists of two subtypes from funnel web spider (*Agelenopsis aperta*) Venom. *J. Biol. Chem.* **265**, 861–867.

Adams, M. E., Herold, E. E. and Venema, V. J. (1989). Two classes of channel-specific toxins from funnel web spider venom. *J. Comp. Physiol. A* **164**, 333–342.

Adams, M. E., Mintz, I. M., Reily, M. D., Thanabal, V. and Bean, B. P. (1993). Structure and properties of ω-agatoxin IVB, a new antagonist of P-type calcium channels. *Mol. Pharmacol.* **44**, 681–688.

Akhunov, A. A., Makevnina, L. G., Golubenko, Z. and Paskhina, T. S. (1996). Kininase of the *Latrodectus tredecimguttatus* venom: a study of its enzyme substrate specificity. *Immunopharmacology* **32**, 160–162.

Ascenzi, P., Bocedi, A., Bolognesi, M., Spallarossa, A., Coletta, M., De Cristofaro, R. and Menegatti, E. (2003). The bovine basic pancreatic trypsin inhibitor (Kunitz inhibitor): a milestone protein. *Curr. Protein Pept. Sci.* **4**, 231–251.

Atkinson, R. K. and Wright, L. G. (1992). The involvement of collagenase in the necrosis induced by the bites of some spiders. *Comp. Biochem. Physiol. C* **102**, 125–128.

Barbaro, K. C., Knysak, I., Martins, R., Hogan, C. and Winkel, K. (2005). Enzymatic characterization, antigenic cross-reactivity and neutralization of dermonecrotic activity of five *Loxosceles* spider venoms of medical importance in the Americas. *Toxicon* **45**, 489–499.

Baumann, T., Kämpfer, U., Schürch, S., Schaller, J., Largiadèr, C., Nentwig, W. and Kuhn-Nentwig, L. (2010a). Ctenidins: antimicrobial glycine-rich peptides from the hemocytes of the spider *Cupiennius salei*. *Cell. Mol. Life Sci.* **67**, 2787–2798.

Baumann, T., Kuhn-Nentwig, L., Largiadèr, C. R. and Nentwig, W. (2010b). Expression of defensins in non-infected araneomorph spiders. *Cell. Mol. Life Sci.* **67**, 2643–2651.

Beleboni, R. O., Guizzo, R., Fontana, A. C., Pizzo, A. B., Carolino, R. O., Gobbo-Neto, L., Lopes, N. P., Coutinho-Netto, J. and Dos Santos, W. F. (2006). Neurochemical characterization of a neuroprotective compound from *Parawixia bistriata* spider venom that inhibits synaptosomal uptake of GABA and glycine. *Mol. Pharmacol.* **69**, 1998–2006.

Benli, M. and Yigit, N. (2008). Antibacterial activity of venom from funnel web spider *Agelena labyrinthica* (Araneae: Agelenidae). *J. Venom. Anim. Toxins* **14**(4).

Berndt, K. D., Guntert, P., Orbons, L. P. and Wüthrich, K. (1992). Determination of a high-quality nuclear magnetic resonance solution structure of the bovine pancreatic trypsin inhibitor and comparison with three crystal structures. *J. Mol. Biol.* **227**, 757–775.

Bettini, S. (1978). Arthropod Venoms. Handbook of Experimental Pharmacology. Springer, Berlin.

Bettini, S. and Maroli, M. (1978). Venoms of Theridiidae, genus *Latrodectus*. In: *Arthropod Venoms. Handbook of Experimental Pharmacology* (ed Bettini, S.), pp. 149–216. Springer, Berlin.

Biass, D., Dutertre, S., Gerbault, A., Menou, J. L., Offord, R., Favreau, P. and Stöcklin, R. (2009). Comparative proteomic study of the venom of the piscivorous cone snail *Conus consors*. *J. Proteomics* **72**, 210–218.

Biel, M. (2009). Cyclic nucleotide-regulated cation channels. *J. Biol. Chem.* **284**, 9017–9021.

Bindokas, V. P. and Adams, M. E. (1989). ω-Aga-I: a presynaptic calcium channel antagonist from venom of the funnel web spider, *Agelenopsis aperta*. *J. Neurobiol.* **20**, 171–188.
Binford, G. J. (2001). Differences in venom composition between orb-weaving and wandering Hawaiian *Tetragnatha* (Araneae). *Biol. J. Linn. Soc.* **74**, 581–595.
Binford, G. J., Bodner, M. R., Cordes, M. H., Baldwin, K. L., Rynerson, M. R., Burns, S. N. and Zobel-Thropp, P. A. (2009). Molecular evolution, functional variation, and proposed nomenclature of the gene family that includes sphingomyelinase D in sicariid spider venoms. *Mol. Biol. Evol.* **26**, 547–566.
Binford, G. J., Callahan, M. S., Bodner, M. R., Rynerson, M. R., Nunez, P. B., Ellison, C. E. and Duncan, R. P. (2008). Phylogenetic relationships of *Loxosceles* and *Sicarius* spiders are consistent with Western Gondwanan vicariance. *Mol. Phylogenet. Evol.* **49**, 538–553.
Binford, G. J. and Wells, M. A. (2003). The phylogenetic distribution of sphingomyelinase D activity in venoms of haplogyne spiders. *Comp. Biochem. Physiol. B Biochem. Mol. Biol.* **135**, 25–33.
Bingham, R. J., Dive, V., Phillips, S. E. V., Shirras, A. D. and Isaac, R. E. (2006). Structural diversity of angiotensin-converting enzyme. *FEBS J.* **273**, 362–373.
Blenau, W. and Baumann, A. (2001). Molecular and pharmacological properties of insect biogenic amine receptors: lessons from *Drosophila melanogaster* and *Apis mellifera*. *Arch. Insect Biochem. Physiol.* **48**, 13–38.
Boevé, J. L., Kuhn-Nentwig, L., Keller, S. and Nentwig, W. (1995). Quantity and quality of venom released by a spider (*Cupiennius salei*, Ctenidae). *Toxicon* **33**, 1347–1357.
Bohlen, C. J., Priel, A., Zhou, S., King, D., Siemens, J. and Julius, D. (2010). A bivalent tarantula toxin activates the capsaicin receptor, TRPV1, by targeting the outer pore domain. *Cell* **141**, 834–845.
Borges, M. H., de Lima, M. E., Stankiewicz, M., Pelhate, M., Cordeiro, M. N. and Beirão, P. S. (2009). Structural and functional diversity in the venom of spiders of the genus *Phoneutria*. In: *Animal Toxins: State of the Art* (ed de Lima, M.), pp. 291–311. Editora UFMG, Belo Horizonte.
Bosmans, F., Escoubas, P. and Nicholson, G. M. (2009). Spider venom peptides as leads for drug and insecticide design. In: *Animal Toxins: State of the Art* (ed de Lima, M.), pp. 269–290. Editora UFMG, Belo Horizonte.
Bosmans, F., Rash, L., Zhu, S., Diochot, S., Lazdunski, M., Escoubas, P. and Tytgat, J. (2006). Four novel tarantula toxins as selective modulators of voltage-gated sodium channel subtypes. *Mol. Pharmacol.* **69**, 419–429.
Bosmans, F. and Swartz, K. J. (2010). Targeting voltage sensors in sodium channels with spider toxins. *Trends Pharmacol. Sci.* **31**, 175–182.
Bowman, C. L., Gottlieb, P. A., Suchyna, T. M., Murphy, Y. K. and Sachs, F. (2007). Mechanosensitive ion channels and the peptide inhibitor GsMTx-4: history, properties, mechanisms and pharmacology. *Toxicon* **49**, 249–270.
Branton, W. D., Rudnick, M. S., Zhou, Y., Eccleston, E. D., Fields, G. B. and Bowers, L. D. (1993). Fatty acylated toxin structure. *Nature* **365**, 496–497.
Brogden, K. A. (2005). Antimicrobial peptides: pore formers or metabolic inhibitors in bacteria? *Nat. Rev. Microbiol.* **3**, 238–250.
Buchner, E., Buchner, S., Burg, M. G., Hofbauer, A., Pak, W. L. and Pollack, I. (1993). Histamine is a major mechanosensory neurotransmitter candidate in *Drosophila melanogaster*. *Cell Tissue Res.* **273**, 119–125.
Budd, T., Clinton, P., Dell, A., Duce, I. R., Johnson, S. J., Quicke, D. L. J., Taylor, G. W., Usherwood, P. N. R. and Usoh, G. (1988). Isolation and characterization of glutamate receptor antagonists from venoms of orb-web spiders. *Brain Res.* **448**, 30–39.

Budnik, B. A., Olsen, J. V., Egorov, T. A., Anisimova, V. E., Galkina, T. G., Musolyamov, A. K., Grishin, E. V. and Zubarev, R. A. (2004). De novo sequencing of antimicrobial peptides isolated from the venom glands of the wolf spider *Lycosa singoriensis*. *J. Mass Spectrom.* **39**, 193–201.

Bulet, P. and Stöcklin, R. (2005). Insect antimicrobial peptides: structures, properties and gene regulation. *Protein Pept. Lett.* **12**, 3–11.

Bulet, P., Stöcklin, R. and Menin, L. (2004). Anti-microbial peptides: from invertebrates to vertebrates. *Immunol. Rev.* **198**, 169–184.

Cabbiness, S. G., Gehrke, C. W., Kuo, K. C., Chan, T. K., Hall, J. E., Hudiburg, S. A. and Odell, G. V. (1980). Polyamines in some tarantula venoms. *Toxicon* **18**, 681–683.

Cavalieri, M., D'Urso, D., Lassa, A., Pierdominici, E., Robello, M. and Grasso, A. (1987). Characterization and some properties of the venom gland extract of a theridiid spider (*Steatoda paykulliana*) frequently mistaken for black widow spider (*Latrodectus tredecimguttatus*). *Toxicon* **25**, 965–974.

Cesar, L. M., Mendes, M. A., Tormena, C. F., Marques, M. R., de Souza, B. M., Saidemberg, D. M., Bittencourt, J. C. and Palma, M. S. (2005a). Isolation and chemical characterization of PwTx-II: a novel alkaloid toxin from the venom of the spider *Parawixia bistriata* (Araneidae, Araneae). *Toxicon* **46**, 786–796.

Cesar, L. M. M., Tormena, C. F., Marques, M. R., Silva, G. V. J., Mendes, M. A., Rittner, R. and Palma, M. S. (2005b). Structure determination of hydroxytrypargine: a new tetrahydro-β-carboline toxin from the venom of the spider *Parawixia bistriata*. *Helv. Chim. Acta* **88**, 796–801.

Chan, T. K., Geren, C. R., Howell, D. E. and Odell, G. V. (1975). Adenosine triphosphate in tarantula spider venoms and its synergistic effect with the venom toxin. *Toxicon* **13**, 61–66.

Chaves-Moreira, D., Chaim, O. M., Sade, Y. B., Paludo, K. S., Gremski, L. H., Donatti, L., de Moura, J., Mangili, O. C., Gremski, W., da Silveira, R. B., Senff-Ribeiro, A. and Veiga, S. S. (2009). Identification of a direct hemolytic effect dependent on the catalytic activity induced by phospholipase-D (dermonecrotic toxin) from brown spider venom. *J. Cell. Biochem.* **107**, 655–666.

Chen, J., Deng, M., He, Q., Meng, E., Jiang, L., Liao, Z., Rong, M. and Liang, S. (2008a). Molecular diversity and evolution of cystine knot toxins of the tarantula *Chilobrachys jingzhao*. *Cell. Mol. Life Sci.* **65**, 2431–2444.

Chen, J., Zhao, L., Jiang, L., Meng, E., Zhang, Y., Xiong, X. and Liang, S. (2008b). Transcriptome analysis revealed novel possible venom components and cellular processes of the tarantula *Chilobrachys jingzhao* venom gland. *Toxicon* **52**, 794–806.

Chesnov, S., Bigler, L. and Hesse, M. (2001). The acylpolyamines from the venom of the spider *Agelenopsis aperta*. *Helv. Chim. Acta* **84**, 2178–2197.

Choi, S. J., Parent, R., Guillaume, C., Deregnaucourt, C., Delarbre, C., Ojcius, D. M., Montagne, J. J., Celerier, M. L., Phelipot, A., Amiche, M., Molgo, J., Camadro, J. M. and Guette, C. (2004). Isolation and characterization of Psalmopeotoxin I and II: two novel antimalarial peptides from the venom of the tarantula *Psalmopoeus cambridgei*. *FEBS Lett.* **572**, 109–117.

Chong, Y., Hayes, J. L., Sollod, B., Wen, S., Wilson, D. T., Hains, P. G., Hodgson, W. C., Broady, K. W., King, G. F. and Nicholson, G. M. (2007). The ω-atracotoxins: selective blockers of insect M-LVA and HVA calcium channels. *Biochem. Pharmacol.* **74**, 623–638.

Cohen, E. and Quistad, G. B. (1998). Cytotoxic effects of arthropod venoms on various cultured cells. *Toxicon* **36**, 353–358.

Conde, R., Zamudio, F. Z., Rodriguez, M. H. and Possani, L. D. (2000). Scorpine, an anti-malaria and anti-bacterial agent purified from scorpion venom. *FEBS Lett.* **471**, 165–168.

Cordeiro, M. N., Figueiredo, S. G., Valentim, A. C., Diniz, C. R., Von Eickstedt, V. R. D., Gilroy, J. and Richardson, M. (1993). Purification and amino acid sequences of six Tx3 type neurotoxins from the venom of the Brazilian armed spider *Phoneutria nigriventer* keys. *Toxicon* **31**, 35–42.

Cordes, M. H. and Binford, G. J. (2006). Lateral gene transfer of a dermonecrotic toxin between spiders and bacteria. *Bioinformatics* **22**, 264–268.

Corzo, G. and Escoubas, P. (2003). Pharmacologically active spider peptide toxins. *Cell. Mol. Life Sci.* **60**, 2409–2426.

Corzo, G., Gilles, N., Satake, H., Villegas, E., Dai, L., Nakajima, T. and Haupt, J. (2003). Distinct primary structures of the major peptide toxins from the venom of the spider *Macrothele gigas* that bind to sites 3 and 4 in the sodium channel. *FEBS Lett.* **547**, 43–50.

Corzo, G., Villegas, E., Gomez-Lagunas, F., Possani, L. D., Belokoneva, O. S. and Nakajima, T. (2002). Oxyopinins, large amphipathic peptides isolated from the venom of the wolf spider *Oxyopes kitabensis* with cytolytic properties and positive insecticidal cooperativity with spider neurotoxins. *J. Biol. Chem.* **277**, 23627–23637.

Cromer, B. A. and McIntyre, P. (2008). Painful toxins acting at TRPV1. *Toxicon* **51**, 163–173.

Cruz, J. S., Cotta, G., Diniz, C. R. and Beirão, P. S. (1994). Partial purification and pharmacological characterization of a neurotoxic fraction isolated from the venom of the spider *Lycosa erythrognatha*. *Braz. J. Med. Biol. Res.* **27**, 2653–2659.

Cunha, R. B., Barbaro, K. C., Muramatsu, D., Portaro, F. C., Fontes, W. and de Sousa, M. V. (2003). Purification and characterization of loxnecrogin, a dermonecrotic toxin from *Loxosceles gaucho* brown spider venom. *J. Protein Chem.* **22**, 135–146.

Cushing, P. E. and Santangelo, R. G. (2008). Notes on the natural history and hunting behaviour of an ant eating zodariid spider (Arachnida, Araneae) in Colorado. *J. Arachnol.* **30**, 618–621.

da Silveira, R. B., Chaim, O. M., Mangili, O. C., Gremski, W., Dietrich, C. P., Nader, H. B. and Veiga, S. S. (2007a). Hyaluronidases in *Loxosceles intermedia* (brown spider) venom are endo-beta-N-acetyl-d-hexosaminidases hydrolases. *Toxicon* **49**, 758–768.

da Silveira, R. B., dos Santos Filho, J. F., Mangili, O. C., Veiga, S. S., Gremski, W., Nader, H. B. and von Dietrich, C. P. (2002). Identification of proteases in the extract of venom glands from brown spiders. *Toxicon* **40**, 815–822.

da Silveira, R. B., Wille, A. C., Chaim, O. M., Appel, M. H., Silva, D. T., Franco, C. R., Toma, L., Mangili, O. C., Gremski, W., Dietrich, C. P., Nader, H. B. and Veiga, S. S. (2007b). Identification, cloning, expression and functional characterization of an astacin-like metalloprotease toxin from *Loxosceles intermedia* (brown spider) venom. *Biochem. J.* **406**, 355–363.

Dai, L., Corzo, G., Naoki, H., Andriantsiferana, M. and Nakajima, T. (2002). Purification, structure-function analysis, and molecular characterization of novel linear peptides from scorpion *Opisthacanthus madagascariensis*. *Biochem. Biophys. Res. Commun.* **293**, 1514–1522.

de Castro, C. S., Silvestre, F. G., Araujo, S. C., Gabriel de, M. Y., Mangili, O. C., Cruz, I., Chavez-Olortegui, C. and Kalapothakis, E. (2004). Identification and molecular cloning of insecticidal toxins from the venom of the brown spider *Loxosceles intermedia*. *Toxicon* **44**, 273–280.

DeLano, W. L. (2002). The PyMOL Molecular Graphics System. Available at: http://www.pymol.org.

Dennison, S. R., Harris, F., Bhatt, T., Singh, J. and Phoenix, D. A. (2009). The effect of C-terminal amidation on the efficacy and selectivity of antimicrobial and anticancer peptides. *Mol. Cell. Biochem.* **332**, 43–50.

Devaraja, S., Girish, K. S., Devaraj, V. R. and Kemparaju, K. (2010a). Factor Xa-like and fibrin(ogen)olytic activities of a serine protease from *Hippasa agelenoides* spider venom gland extract. *J. Thromb. Thrombolysis* **29**, 119–126.

Devaraja, S., Girish, K. S., Gowtham, Y. J. and Kemparaju, K. (2010b). The hag-protease-II is a fibrin(ogen)ase from *Hippasa agelenoides* spider venom gland extract: purification, characterization and its role in hemostasis. *Toxicon* **57**, 248–258.

Devaraja, S., Nagaraju, S., Mahadeswaraswamy, Y. H., Girish, K. S. and Kemparaju, K. (2008). A low molecular weight serine protease: purification and characterization from *Hippasa agelenoides* (funnel web) spider venom gland extract. *Toxicon* **52**, 130–138.

Diaz, J. H. (2004). The global epidemiology, syndromic classification, management, and prevention of spider bites. *Am. J. Trop. Med. Hyg.* **71**, 239–250.

Diego-Garcia, E., Peigneur, S., Waelkens, E., Debaveye, S. and Tytgat, J. (2010). Venom components from *Citharischius crawshayi* spider (Family Theraphosidae): exploring transcriptome, venomics, and function. *Cell. Mol. Life Sci.* **67**, 2799–2813.

Dippenaar, A. S. and Meyer, M. K. P. (1980). On the species of the African genus *Ammoxenus* (Araneae: Ammoxenidae), with description of two new species. *J. Entomol. Soc. South. Afr.* **43**, 41–49.

Döring, F., Wischmeyer, E., Kuhnlein, R. P., Jackle, H. and Karschin, A. (2002). Inwardly rectifying K^+ (Kir) channels in *Drosophila*. A crucial role of cellular milieu factors Kir channel function. *J. Biol. Chem.* **277**, 25554–25561.

Duffield, P. H., Duffield, A. M., Carroll, P. R. and Morgans, D. (1979). Analysis of the venom of the Sydney funnel-web spider. *Atrax robustus* using gas chromatography mass spectrometry. *Biomed. Mass Spectrom.* **6**, 105–108.

Dutertre, S., Biass, D., Stöcklin, R. and Favreau, P. (2010). Dramatic intraspecimen variations within the injected venom of *Conus consors*: an unsuspected contribution to venom diversity. *Toxicon* **55**, 1453–1462.

Dy, C. Y., Buczek, P., Imperial, J. S., Bulaj, G. and Horvath, M. P. (2006). Structure of conkunitzin-S1, a neurotoxin and Kunitz-fold disulfide variant from cone snail. *Acta Crystallogr. D Biol. Crystallogr.* **62**, 980–990.

Early, S. L. and Michaelis, E. K. (1987). Presence of proteins and glutamate as major constituents of the venom of the spider *Araneus gemma*. *Toxicon* **25**, 433–442.

El Hassani, A. K., Giurfa, M., Gauthier, M. and Armengaud, C. (2008). Inhibitory neurotransmission and olfactory memory in honeybees. *Neurobiol. Learn. Mem.* **90**, 589–595.

Escoubas, P., Célérier, M. L., Romi-Lebrun, R. and Nakajima, T. (1997). Two novel peptide neurotoxins from the venom on the tarantula *Lasiodora parahybana*. *Toxicon* **35**, 805–806.

Escoubas, P., De Weille, J. R., Lecoq, A., Diochot, S., Waldmann, R., Champigny, G., Moinier, D., Menez, A. and Lazdunski, M. (2000a). Isolation of a tarantula toxin specific for a class of proton-gated Na^+ channels. *J. Biol. Chem.* **275**, 25116–25121.

Escoubas, P., Diochot, S. and Corzo, G. (2000b). Structure and pharmacology of spider venom neurotoxins. *Biochimie* **82**, 893–907.

Escoubas, P. and King, G. F. (2009). Venomics as a drug discovery platform. *Expert Rev. Proteomics* **6**, 221–224.

Escoubas, P. and Rash, L. (2004). Tarantulas: eight-legged pharmacists and combinatorial chemists. *Toxicon* **43**, 555–574.

Estrada, G., Villegas, E. and Corzo, G. (2007). Spider venoms: a rich source of acylpolyamines and peptides as new leads for CNS drugs. *Nat. Prod. Rep.* **24**, 145–161.

Favreau, P., Gilles, N., Lamthanh, H., Bournaud, R., Shimahara, T., Bouet, F., Laboute, P., Letourneux, Y., Menez, A., Molgo, J. and Le Gall, F. (2001). A new omega-conotoxin that targets N-type voltage-sensitive calcium channels with unusual specificity. *Biochemistry* **40**, 14567–14575.

Feitosa, L., Gremski, W., Veiga, S. S., Elias, M. C. Q. B., Graner, E., Mangili, O. C. and Brentani, R. R. (1998). Detection and characterization of metalloproteinases with gelatinolytic, fibronectinolytic and fibrinogenolytic activities in brown spider (*Loxosceles intermedia*) venom. *Toxicon* **36**, 1039–1051.

Fenton, A. W., West, P. R., Odell, G. V., Hudiburg, S. M., Ownby, C. L., Mills, J. N., Scroggins, B. T. and Shannon, S. B. (1995). Arthropod venom citrate inhibits phospholipase A_2. *Toxicon* **33**, 763–770.

Fernandes-Pedrosa, M. F., Junqueira-de-Azevedo, I. L., Goncalves-de-Andrade, R. M., Kobashi, L. S., Almeida, D. D., Ho, P. L. and Tambourgi, D. V. (2008). Transcriptome analysis of *Loxosceles laeta* (Araneae, Sicariidae) spider venomous gland using expressed sequence tags. *BMC Genomics* **9**, 279.

Ferreira, L. A., Lucas, S. M., Alves, E. W., Hermann, V. V., Reichl, A. P., Habermehl, G. and Zingali, R. B. (1998). Isolation, characterization and biological properties of two kinin-like peptides (peptide-S and peptide-R) from *Scaptocosa raptoria* venom. *Toxicon* **36**, 31–39.

Ferreira, S. H. and Rocha-e-Silva, T. A. A. (1965). Potentiation of bradykinin and eledoisin by BPF (bradykinin potentiating factor) from *Bothrops jararaca* venom. *Experientia* **21**, 347–349.

Fischer, F. G. and Bohn, H. (1957). Die Giftsekrete der brasilianischen Tarantel *Lycosa erythrognatha* und der Wanderspinne *Phoneutria fera*. *Hoppe-Seyler's Z. Physiol. Chem.* **306**, 265–268.

Fisyunov, A., Pluzhnikov, K., Molyavka, A., Grishin, E., Lozovaya, N. and Krishtal, O. (2005). Novel spider toxin slows down the activation kinetics of P-type Ca^{2+} channels in Purkinje neurons of rat. *Toxicology* **207**, 129–136.

Foelix, R. and Erb, B. (2010). Mesothelae have venom glands. *J. Arachnol.* **38**, 596–598.

Foelix, R. F. (2011). Biology of Spiders. Oxford University Press, New York, USA.

Foradori, M. J., Smith, S. C., Smith, E. and Wells, R. E. (2005). Survey for potentially necrotizing spider venoms, with special emphasis on *Cheiracanthium mildei*. *Comp. Biochem. Physiol. C Toxicol. Pharmacol.* **141**, 32–39.

Forster, R. R. and Platnick, N. I. (1977). A review of the spider family Symphytognathidae (Arachnida, Araneae). *Am. Mus. Novit.* **2619**, 1–29.

Forsythe, I. D. (1995). A physiological function for polyamines? *Curr. Biol.* **5**, 1248–1251.

Frew, R., Hamilton, M. G. and Lundy, P. M. (1994). Identification of noradrenaline in venom from the funnel-web spider *Hololena curta*. *Toxicon* **32**, 511–515.

Friedel, T. and Nentwig, W. (1989). Immobilizing and lethal effects of spider venoms on the cockroach and the common mealbeetle. *Toxicon* **27**, 305–316.

Fry, B. G., Roelants, K., Champagne, D. E., Scheib, H., Tyndall, J. D., King, G. F., Nevalainen, T. J., Norman, J. A., Lewis, R. J., Norton, R. S., Renjifo, C. and de la Vega, R. C. (2009). The toxicogenomic multiverse: convergent recruitment of proteins into animal venoms. *Annu. Rev. Genomics Hum. Genet.* **10**, 483–511.

Gasparini, S., Kiyatkin, N., Drevet, P., Boulain, J. C., Tacnet, F., Ripoche, P., Forest, E., Grishin, E. and Menez, A. (1994). The low molecular weight protein which copurifies with alpha-latrotoxin is structurally related to crustacean hyperglycemic hormones. *J. Biol. Chem.* **269**, 19803–19809.

Gelfuso, E. A., Cunha, A. O., Mortari, M. R., Liberato, J. L., Paraventi, K. H., Beleboni, R. O., Coutinho-Netto, J., Lopes, N. P. and dos Santos, W. F. (2007).

Neuropharmacological profile of FrPbAII, purified from the venom of the social spider *Parawixia bistriata* (Araneae, Araneidae), in Wistar rats. *Life Sci.* **80**, 566–572.

Gelly, J. C., Gracy, J., Kaas, Q., Le-Nguyen, D., Heitz, A. and Chiche, L. (2004). The KNOTTIN website and database: a new information system dedicated to the knottin scaffold. *Nucleic Acids Res.* **32**, D156–D159.

Gemeno, C., Yeargan, K. V. and Haynes, K. F. (2000). Aggressive chemical mimicry by the bolas spider *Mastophora hutchinsoni*: identification and quantification of a major prey's sex pheromone components in the spider's volatile emissions. *J. Chem. Ecol.* **26**, 1235–1243.

Geren, C. R., Chan, T. K., Howell, D. E. and Odell, G. V. (1976). Isolation and characterization of toxins from brown recluse spider venom (*Loxosceles reclusa*). *Arch. Biochem. Biophys.* **174**, 90–99.

Giangaspero, A., Sandri, L. and Tossi, A. (2001). Amphipathic α-helical antimicrobial peptides – a systematic study of the effects of structural and physical properties on biological activity. *Eur. J. Biochem.* **268**, 5589–5600.

Gisselmann, G., Pusch, H., Hovemann, B. T. and Hatt, H. (2002). Two cDNAs coding for histamine-gated ion channels in *D. melanogaster*. *Nat. Neurosci.* **5**, 11–12.

Goldin, A. L. (1999). Diversity of mammalian voltage-gated sodium channels. *Ann. N. Y. Acad. Sci.* **868**, 38–50.

Gomez, M. V., Kalapothakis, E., Guatimosim, C. and Prado, M. A. (2002). *Phoneutria nigriventer* venom: a cocktail of toxins that affect ion channels. *Cell. Mol. Neurobiol.* **22**, 579–588.

Gracy, J., Le-Nguyen, D., Gelly, J. C., Kaas, Q., Heitz, A. and Chiche, L. (2008). KNOTTIN: the knottin or inhibitor cystine knot scaffold in 2007. *Nucleic Acids Res.* **36**, D314–D319.

Grasso, A. and Pescatori, M. (1996). Structural and functional studies of latrodectin from the venom of black widow spider (*Latrodectus tredecimguttatus*). *Adv. Exp. Med. Biol.* **391**, 237–243.

Graudins, A., Gunja, N., Broady, K. W. and Nicholson, G. M. (2002a). Clinical and in vitro evidence for the efficacy of Australian red-back spider (*Latrodectus hasselti*) antivenom in the treatment of envenomation by a cupboard spider (*Steatoda grossa*). *Toxicon* **40**, 767–775.

Graudins, A., Wilson, D., Alewood, P. F., Broady, K. W. and Nicholson, G. M. (2002b). Cross-reactivity of Sydney funnel-web spider antivenom: neutralization of the in vitro toxicity of other Australian funnel-web (*Atrax* and *Hadronyche*) spider venoms. *Toxicon* **40**, 259–266.

Gremski, L. H., da Silveira, R. B., Chaim, O. M., Probst, C. M., Ferrer, V. P., Nowatzki, J., Weinschutz, H. C., Madeira, H. M., Gremski, W., Nader, H. B., Senff-Ribeiro, A. and Veiga, S. S. (2010). A novel expression profile of the *Loxosceles intermedia* spider venomous gland revealed by transcriptome analysis. *Mol. Biosyst.* **6**, 2403–2416.

Grishin, E. V. (1998). Black widow spider toxins: the present and the future. *Toxicon* **36**, 1693–1701.

Grishin, E. V., Volkova, T. M. and Arseniev, A. S. (1989). Isolation and structure analysis of components from venom of the spider *Argiope lobata*. *Toxicon* **27**, 541–549.

Gunasekaran, K., Nagarajaram, H. A., Ramakrishnan, C. and Balaram, P. (1998). Stereochemical punctuation marks in protein structures: glycine and proline containing helix stop signals. *J. Mol. Biol.* **275**, 917–932.

Gunning, S. J., Maggio, F., Windley, M. J., Valenzuela, S. M., King, G. F. and Nicholson, G. M. (2008). The Janus-faced atracotoxins are specific blockers of invertebrate K_{Ca} channels. *FEBS J.* **275**, 4045–4059.

Guo, D. and Lu, Z. (2000). Mechanism of cGMP-gated channel block by intracellular polyamines. *J. Gen. Physiol.* **115**, 783–798.

Hardie, R. C. (1987). Is histamine a neurotransmitter in insect photoreceptors? *J. Comp. Physiol. A* **161**, 201–213.

Hardie, R. C. (1989). A histamine-activated chloride channel involved in neurotransmission at a photoreceptor synapse. *Nature* **339**, 704–706.

Heck, S. D., Siok, C. J., Krapcho, K. J., Kelbaugh, P. R., Thadeio, P. F., Welch, M. J., Williams, R. D., Ganong, A. H., Kelly, M. E., Lanzetti, A. J., Gray, W. R. Phillips, D., *et al.* (1994). Functional consequences of posttranslational isomerization of Ser-46 in a calcium channel toxin. *Science* **266**, 1065–1068.

Heitz, J. R. and Norment, B. R. (1974). Characteristics of an alkaline phosphatase activity in brown recluse venom. *Toxicon* **12**, 181–187.

Herzig, V., Wood, D. L., Newell, F., Chaumeil, P. A., Kaas, Q., Binford, G. J., Nicholson, G. M., Gorse, D. and King, G. F. (2011). ArachnoServer 2.0, an updated online resource for spider toxin sequences and structures. *Nucleic Acids Res.* **39**, D653–D657.

Homberg, U. (2002). Neurotransmitters and neuropeptides in the brain of the locust. *Microsc. Res. Tech.* **56**, 189–209.

Horni, A., Weickmann, D. and Hesse, M. (2001). The main products of the low molecular mass fraction in the venom of the spider *Latrodectus menavodi*. *Toxicon* **39**, 425–428.

Hostettler, S. and Nentwig, W. (2006). Olfactory information saves venom during prey-capture of the hunting spider *Cupiennius salei* (Araneae: Ctenidae). *Funct. Ecol.* **20**, 369–375.

Hurst, A. C., Gottlieb, P. A. and Martinac, B. (2009). Concentration dependent effect of GsMTx4 on mechanosensitive channels of small conductance in *E. coli* spheroplasts. *Eur. Biophys. J.* **38**, 415–425.

Inceoglu, B., Lango, J., Jing, J., Chen, L., Doymaz, F., Pessah, I. N. and Hammock, B. D. (2003). One scorpion, two venoms: prevenom of *Parabuthus transvaalicus* acts as an alternative type of venom with distinct mechanism of action. *Proc. Natl. Acad. Sci. USA* **100**, 922–927.

Iovchev, M., Boutanaev, A., Ivanov, I., Wolstenholme, A., Nurminsky, D. and Semenov, E. (2006). Phylogenetic shadowing of a histamine-gated chloride channel involved in insect vision. *Insect Biochem. Mol. Biol.* **36**, 10–17.

Isbister, G. K. and Gray, M. R. (2002). A prospective study of 750 definite spider bites, with expert spider identification. *QJM* **95**, 723–731.

Isbister, G. K. and Gray, M. R. (2003). Effects of envenoming by comb-footed spiders of the genera *Steatoda* and *Achaearanea* (family Theridiidae: Araneae) in Australia. *J. Toxicol. Clin. Toxicol.* **41**, 809–819.

Ishii, T. M., Nakashima, N., Takatsuka, K. and Ohmori, H. (2007). Peripheral N- and C-terminal domains determine deactivation kinetics of HCN channels. *Biochem. Biophys. Res. Commun.* **359**, 592–598.

Itagaki, Y., Fujita, T., Naoki, H., Yasuhara, T., Andriantsiferana, M. and Nakajima, T. (1997). Detection of new spider toxins from a *Nephilengys borbonica* venom gland using on-line μ-column HPLC continuous flow (FRIT) FAB LC/MS and MS/MS. *Nat. Toxins* **5**, 1–13.

Itagaki, Y. and Nakajima, T. (2000). Acylpolyamines: mass spectrometric analytical methods for Araneidae spider acylpolyamines. *J. Toxicol. Toxin Rev.* **19**, 23–52.

Janssen, D., Derst, C., Rigo, J. M. and Van Kerkhove, E. (2010). Cys-loop ligand-gated chloride channels in dorsal unpaired median neurons of *Locusta migratoria*. *J. Neurophysiol.* **103**, 2587–2598.

Jasys, V. J., Kelbaugh, P. R., Nason, D. M., Phillips, D., Rosnack, K. J., Saccomano, N. A., Stroh, J. G. and Volkmann, R. A. (1990). Isolation, structure elucidation, and synthesis of novel hydroxylamine-containing polyamines from the venom of the *Agelenopsis aperta* spider. *J. Am. Chem. Soc.* **112**, 6696–6704.

Jiang, L., Peng, L., Chen, J., Zhang, Y., Xiong, X. and Liang, S. (2008). Molecular diversification based on analysis of expressed sequence tags from the venom glands of the Chinese bird spider *Ornithoctonus huwena*. *Toxicon* **51**, 1479–1489.

Jiang, L., Peng, L., Zhang, Y., Chen, J., Zhang, D. and Liang, S. (2009). Expression, purification and characterization of a group of lectin-like peptides from the spider *Ornithoctonus huwena*. *Peptides* **30**, 669–674.

Jiang, L., Zhang, D., Zhang, Y., Peng, L., Chen, J. and Liang, S. (2010). Venomics of the spider *Ornithoctonus huwena* based on transcriptomic versus proteomic analysis. *Comp. Biochem. Physiol. Part D Genomics Proteomics* **5**, 81–88.

Jocqué, R. and Dippenaar-Schoeman, A. S. (2006). Spider Families of the World. Musée Royal de L'Afrique Centrale, Tervuren.

Jones, A. K., Bera, A. N., Lees, K. and Sattelle, D. B. (2010). The cys-loop ligand-gated ion channel gene superfamily of the parasitoid wasp, *Nasonia vitripennis*. *Heredity* **104**, 247–259.

Jung, E. H., Lee, K. S., Han, J. H., Je, Y. H., Chang, J. H., Roh, J. Y., Sohn, H. D. and Jin, B. R. (2002). Molecular cloning of two cDNAs encoding an insecticidal toxin from the spider, *Araneus ventricosus*, and construction of a recombinant baculovirus expressing a spider toxin. *Int. J. Indust. Entomol.* **4**, 43–49.

Jung, H. J., Kim, P. I., Lee, S. K., Lee, C. W., Eu, Y. J., Lee, D. G., Earm, Y. E. and Kim, J. I. (2006). Lipid membrane interaction and antimicrobial activity of GsMTx-4, an inhibitor of mechanosensitive channel. *Biochem. Biophys. Res. Commun.* **340**, 633–638.

Kaestner, A. (1969). Lehrbuch der speziellen Zoologie. Fischer, Stuttgart.

Kaiser, E. (1956). Enzymatic activity of spider venoms. *In* Venoms pp. 91–93. American Association for the Advancement of Science, Washington, DC.

Kalapothakis, E., Chatzaki, M., Goncalves-Dornelas, H., de Castro, C. S., Silvestre, F. G., Laborne, F. V., de Moura, J. F., Veiga, S. S., Chavez-Olortegui, C., Granier, C. and Barbaro, K. C. (2007). The Loxtox protein family in *Loxosceles intermedia* (Mello-Leitao) venom. *Toxicon* **50**, 938–946.

Kamolkijkarn, P., Prasertdee, T., Netirojjanakul, C., Sarnpitak, P., Ruchirawat, S. and Deechongkit, S. (2010). Synthesis, biophysical, and biological studies of wild-type and mutant psalmopeotoxins–anti-malarial cysteine knot peptides from *Psalmopoeus cambridgei*. *Peptides* **31**, 533–540.

Kawai, N., Abe, T. and Niwa, A. (1982). Spider venom contains specific receptor blocker of glutaminergic synapses. *Brain Res.* **247**, 169–171.

King, G. F., Escoubas, P. and Nicholson, G. M. (2008). Peptide toxins that selectively target insect Na_V and Ca_V channels. *Channels* **2**, 100–116.

Kitaguchi, T. and Swartz, K. J. (2005). An inhibitor of TRPV1 channels isolated from funnel web spider venom. *Biochemistry* **44**, 15544–15549.

Kiyatkin, N. I., Kulikovskaya, I. M., Grishin, E. V., Beadle, D. J. and King, L. A. (1995). Functional characterization of black widow spider neurotoxins synthesised in insect cells. *Eur. J. Biochem.* **230**, 854–859.

Kostrewa, D., Wyss, M., D'Arcy, A. and van Loon, A. P. (1999). Crystal structure of *Aspergillus niger* pH 2.5 acid phosphatase at 2.4 Å resolution. *J. Mol. Biol.* **288**, 965–974.

Kovoor, J. (1987). Comparative structure and histochemistry of silk-producing organs in arachnids. In: *Ecophysiology of Spiders* (ed Nentwig, W.), pp. 160–186. Springer, Berlin.

Kozlov, S. and Grishin, E. (2005). Classification of spider neurotoxins using structural motifs by primary structure features. Single residue distribution analysis and pattern analysis techniques. *Toxicon* **46**, 672–686.

Kozlov, S., Malyavka, A., McCutchen, B., Lu, A., Schepers, E., Herrmann, R. and Grishin, E. (2005). A novel strategy for the identification of toxin-like structures in spider venom. *Proteins* **59**, 131–140.

Kozlov, S. A., Vassilevski, A. A., Feofanov, A. V., Surovoy, A. Y., Karpunin, D. V. and Grishin, E. V. (2006). Latarcins, antimicrobial and cytolytic peptides from the venom of the spider *Lachesana tarabaevi* (Zodariidae) that exemplify biomolecular diversity. *J. Biol. Chem.* **281**, 20983–20992.

Kubista, H., Mafra, R. A., Chong, Y., Nicholson, G. M., Beirão, P. S., Cruz, J. S., Boehm, S., Nentwig, W. and Kuhn-Nentwig, L. (2007). CSTX-1, a toxin from the venom of the hunting spider *Cupiennius salei*, is a selective blocker of L-type calcium channels in mammalian neurons. *Neuropharmacology* **52**, 1650–1662.

Kuhn-Nentwig, L. (2003). Antimicrobial and cytolytic peptides of venomous arthropods. *Cell. Mol. Life Sci.* **60**, 2651–2668.

Kuhn-Nentwig, L. (2009). Cytolytic and antimicrobial peptides in the venom of scorpions and spiders. In: *Animal Toxins: State of the Art* (ed de Lima, M. E.), pp. 249–268. Editora UFMG, Belo Horizonte.

Kuhn-Nentwig, L., Bücheler, A., Studer, A. and Nentwig, W. (1998). Taurine and histamine: low molecular compounds in prey hemolymph increase the killing power of spider venom. *Naturwissenschaften* **85**, 136–138.

Kuhn-Nentwig, L., Dathe, M., Walz, A., Schaller, J. and Nentwig, W. (2002a). Cupiennin 1d*: the cytolytic activity depends on the hydrophobic N-terminus and is modulated by the polar C-terminus. *FEBS Lett.* **527**, 193–198.

Kuhn-Nentwig, L., Müller, J., Schaller, J., Walz, A., Dathe, M. and Nentwig, W. (2002b). Cupiennin 1, a new family of highly basic antimicrobial peptides in the venom of the spider *Cupiennius salei* (Ctenidae). *J. Biol. Chem.* **277**, 11208–11216.

Kuhn-Nentwig, L., Schaller, J., Kämpfer, U., Imboden, H., Malli, H. and Nentwig, W. (2000). A lysine rich C-terminal tail is directly involved in the toxicity of CSTX-1, a neurotoxic peptide from the venom of the spider *Cupiennius salei*. *Arch. Insect Biochem. Physiol.* **44**, 101–111.

Kuhn-Nentwig, L., Schaller, J. and Nentwig, W. (1994). Purification of toxic peptides and the amino acid sequence of CSTX-1 from the multicomponent venom of *Cupiennius salei* (Araneae: Ctenidae). *Toxicon* **32**, 287–302.

Kuhn-Nentwig, L., Schaller, J. and Nentwig, W. (2004). Biochemistry, toxicology and ecology of the venom of the spider *Cupiennius salei* (Ctenidae). *Toxicon* **43**, 543–553.

Kuhn-Nentwig, L., Willems, J., Seebeck, T., Shalaby, T., Kaiser, M. and Nentwig, W. (2011). Cupiennin 1a exhibits a remarkably broad, non-stereospecific cytolytic activity on bacteria, protozoan parasites, insects, and human cancer cells. *Amino Acids* **40**, 69–76.

Kuntner, M., Coddington, J. A. and Hormiga, G. (2008). Phylogeny of extant nephilid orb-weaving spiders (Araneae, Nephilidae): testing morphological and ethological homologies. *Cladistics* **24**, 147–217.

Kurpiewski, G., Forrester, L. J., Barrett, J. T. and Campbell, B. J. (1981). Platelet aggregation and sphingomyelinase D activity of a purified toxin from the venom of *Loxosceles reclusa*. *Biochim. Biophys. Acta* **678**, 467–476.

Lamango, N. S., Sajid, M. and Isaac, R. E. (1996). The endopeptidase activity and the activation by Cl- of angiotensin-converting enzyme is evolutionarily conserved: purification and properties of an an angiotensin-converting enzyme from the housefly, *Musca domestica*. *Biochem. J.* **314**, 639–646.

Lange, C., Paris, C. and Celerier, M. L. (1992). The components of the venom of a spider *Scodra griseipes* 1. Analysis of low molecular weight products using gas chromatography-mass spectrometry. *Rapid Commun. Mass Spectrom.* **6**, 289–292.

Lee, S. and Lynch, K. R. (2005). Brown recluse spider (*Loxosceles reclusa*) venom phospholipase D (PLD) generates lysophosphatidic acid (LPA). *Biochem. J.* **391**, 317–323.

Lee, Y. L., Thrupp, L., Owens, J., Cesario, T. and Shanbrom, E. (2001). Bactericidal activity of citrate against Gram-positive cocci. *Lett. Appl. Microbiol.* **33**, 349–351.

Lewis, R. J. and Garcia, M. L. (2003). Therapeutic potential of venom peptides. *Nat. Rev. Drug Discov.* **2**, 790–802.

Liang, S. (2004). An overview of peptide toxins from the venom of the Chinese bird spider *Selenocosmia huwena* Wang [=*Ornithoctonus huwena* (Wang)]. *Toxicon* **43**, 575–585.

Liang, S. (2008). Proteome and peptidome profiling of spider venoms. *Expert Rev. Proteomics* **5**, 731–746.

Liao, Z., Cao, J., Li, S., Yan, X., Hu, W., He, Q., Chen, J., Tang, J., Xie, J. and Liang, S. (2007). Proteomic and peptidomic analysis of the venom from Chinese tarantula *Chilobrachys jingzhao*. *Proteomics* **7**, 1892–1907.

Liao, Z., Yuan, C., Deng, M., Li, J., Chen, J., Yang, Y., Hu, W. and Liang, S. (2006). Solution structure and functional characterization of jingzhaotoxin-XI: a novel gating modifier of both potassium and sodium channels. *Biochemistry* **45**, 15591–15600.

Liberato, J. L., Cunha, A. O., Mortari, M. R., Gelfuso, E. A., Beleboni Rde, O., Coutinho-Netto, J. and dos Santos, W. F. (2006). Anticonvulsant and anxiolytic activity of FrPbAII, a novel GABA uptake inhibitor isolated from the venom of the social spider *Parawixia bistriata* (Araneidae: Araneae). *Brain Res.* **1124**, 19–27.

Lorenzini, D. M., da Silva, P. I. Jr., Fogaca, A. C., Bulet, P. and Daffre, S. (2003). Acanthoscurrin: a novel glycine-rich antimicrobial peptide constitutively expressed in the hemocytes of the spider *Acanthoscurria gomesiana*. *Dev. Comp. Immunol.* **27**, 781–791.

MacLean, S. J., Andrews, B. C. and Verheyen, E. M. (2002). Characterization of Dir: a putative potassium inward rectifying channel in *Drosophila*. *Mech. Dev.* **116**, 193–197.

Malli, H., Kuhn-Nentwig, L., Imboden, H. and Nentwig, W. (1999). Effects of size, motility and paralysation time of prey on the quantity of venom injected by the hunting spider *Cupiennius salei*. *J. Exp. Biol.* **202**, 2083–2089.

Malli, H., Vapenik, Z. and Nentwig, W. (1993). Ontogenetic changes in the toxicity of the spider *Cupiennius salei* (Araneae, Ctenidae). *Zool. Jb. Physiol.* **97**, 113–122.

Margaret, G. R. and Phanuel, G. J. (1988). Preliminary studies on the venom of three Indian spiders. *Proc. Indian Acad. Sci.* **97**, 231–237.

Marques, M. R., Mendes, M. A., Tormena, C. F., Souza, B. M., Cesar, L. M., Rittner, R. and Palma, M. S. (2005). Structure determination of a tetrahydro-beta-carboline of arthropod origin: a novel alkaloid-toxin subclass from the web of spider *Nephila clavipes*. *Chem. Biodivers.* **2**, 525–534.

Marques, M. R., Mendes, M. A., Tormena, C. F., Souza, B. M., Ribeiro, S. P., Rittner, R. and Palma, M. S. (2004). Structure determination of an organometallic 1-(diazenylaryl)ethanol: a novel toxin subclass from the web of the spider *Nephila clavipes*. *Chem. Biodivers.* **1**, 830–838.

Matsuzaki, K. (1998). Magainins as paradigm for the mode of action of pore forming polypeptides. *Biochim. Biophys. Acta* **1376**, 391–400.

Matsuzaki, K. (2009). Control of cell selectivity of antimicrobial peptides. *Biochim. Biophys. Acta* **1788**, 1687–1692.

McCormick, J., Li, Y., McCormick, K., Duynstee, H. I., van Engen, A. K., van der Marel, G. A., Ganem, B., van Boom, J. H. and Meinwald, J. (1999). Structure and total synthesis of HF-7, a neuroactive glyconucleoside disulfate from the funnel-web spider *Hololena curta*. *J. Am. Chem. Soc.* **121**, 5661–5665.

McCormick, K. D. and Meinwald, J. (1993). Neurotoxic acylpolyamines from spider venoms. *J. Chem. Ecol.* **19**, 2411–2451.

Meehan, C. J., Olson, E. J., Reudink, M. W., Kyser, T. K. and Curry, R. L. (2009). Herbivory in a spider through exploitation of an ant–plant mutualism. *Curr. Biol.* **19**, R892–R893.

Mellor, I. R. and Usherwood, P. N. (2004). Targeting ionotropic receptors with polyamine-containing toxins. *Toxicon* **43**, 493–508.

Mendes, M. A. and Palma, M. S. (2006). Two new bradykinin-related peptides from the venom of the social wasp *Protopolybia exigua* (Saussure). *Peptides* **27**, 2632–2639.

Ménez, A., Stöcklin, R. and Mebs, D. (2006). 'Venomics' or: The venomous systems genome project. *Toxicon* **47**, 255–259.

Meyer, W. (1991). The biology of excitatory neurotransmitters in the central nervous system of spiders (Arachnida: Araneida). *Neuro. Chem.* **10**, 85–103.

Minton, S. A. (1974). Venom Diseases. Charles C Thomas Publishing, Illinois.

Miranda, D. M., Romano-Silva, M. A., Kalapothakis, E., Diniz, C. R., Cordeiro, M. N., Santos, T. M., Prado, M. A. and Gomez, M. V. (1998). *Phoneutria nigriventer* toxins block tityustoxin-induced calcium influx in synaptosomes. *Neuroreport* **9**, 1371–1373.

Mironov, S. L., Sokolov Yu, V., Chanturiya, A. N. and Lishko, V. K. (1986). Channels produced by spider venoms in bilayer lipid membrane: mechanisms of ion transport and toxic action. *Biochim. Biophys. Acta* **862**, 185–198.

Miyazu, M., Tanimura, T. and Sokabe, M. (2000). Molecular cloning and characterization of a putative cyclic nucleotide-gated channel from *Drosophila melanogaster*. *Insect Mol. Biol.* **9**, 283–292.

Moerman, L., Bosteels, S., Noppe, W., Willems, J., Clynen, E., Schoofs, L., Thevissen, K., Tytgat, J., Van Eldere, J., Van Der Walt, J. and Verdonck, F. (2002). Antibacterial and antifungal properties of alpha-helical, cationic peptides in the venom of scorpions from southern Africa. *Eur. J. Biochem.* **269**, 4799–4810.

Moore, E. L., Haspel, G., Libersat, F. and Adams, M. E. (2006). Parasitoid wasp sting: a cocktail of GABA, taurine, and beta-alanine opens chloride channels for central synaptic block and transient paralysis of a cockroach host. *J. Neurobiol.* **66**, 811–820.

Moore, S., Smyth, W. F., Gault, V. A., O'Kane, E. and McClean, S. (2009). Mass spectrometric characterisation and quantitation of selected low molecular mass compounds from the venom of *Haplopelma lividum* (Theraphosidae). *Rapid Commun. Mass Spectrom.* **23**, 1747–1755.

Mouhat, S., Andreotti, N., Jouirou, B. and Sabatier, J. M. (2008). Animal toxins acting on voltage-gated potassium channels. *Curr. Pharm. Des.* **14**, 2503–2518.

Nagaraju, S., Devaraja, S. and Kemparaju, K. (2007a). Purification and properties of hyaluronidase from *Hippasa partida* (funnel web spider) venom gland extract. *Toxicon* **50**, 383–393.

Nagaraju, S., Girish, K. S., Fox, J. W. and Kemparaju, K. (2007b). 'Partitagin' a hemorrhagic metalloprotease from *Hippasa partita* spider venom: role in tissue necrosis. *Biochimie* **89**, 1322–1331.

Nagaraju, S., Mahadeswaraswamy, Y. H., Girish, K. S. and Kemparaju, K. (2006). Venom from spiders of the genus *Hippasa*: biochemical and pharmacological studies. *Comp. Biochem. Physiol. C Toxicol. Pharmacol.* **144**, 1–9.

Narahashi, T., Zhao, X., Ikeda, T., Nagata, K. and Yeh, J. Z. (2007). Differential actions of insecticides on target sites: basis for selective toxicity. *Hum. Exp. Toxicol.* **26**, 361–366.

Nässel, D. R. (1999). Histamine in the brain of insects: a review. *Microsc. Res. Tech.* **44**, 121–136.

Nentwig, W. (1985). Social spiders chatch larger prey: a study of *Anelosimus eximius* (Araneae: Theridiidae). *Behav. Ecol. Sociobiol.* **17**, 79–85.

Nentwig, W. (1987). The prey of spiders. In: *Ecophysiology of Spiders* (ed Nentwig, W.), pp. 249–263. Springer, Berlin.

Nentwig, W., Friedel, T. and Manhart, C. (1992). Comparative investigations on the effect of the venom of 18 spider species onto the cockroach *Blatta orientalis* (Blattodea). *Zool. Jb. Physiol.* **96**, 279–290.

Nentwig, W. and Wissel, C. (1986). A comparison of prey length among spiders. *Oecologia* **68**, 595–600.

Nicholson, G. M. (2007). Insect-selective spider toxins targeting voltage-gated sodium channels. *Toxicon* **49**, 490–512.

Nicholson, G. M., Graudins, A., Wilson, H. I., Little, M. and Broady, K. W. (2006). Arachnid toxinology in Australia: from clinical toxicology to potential applications. *Toxicon* **48**, 872–898.

Nicholson, G. M., Little, M. J. and Birinyi-Strachan, L. C. (2004). Structure and function of δ-atracotoxins: lethal neurotoxins targeting the voltage-gated sodium channel. *Toxicon* **43**, 587–599.

Nishizawa, M. and Nishizawa, K. (2007). Molecular dynamics simulations of a stretch-activated channel inhibitor GsMTx4 with lipid membranes: two binding modes and effects of lipid structure. *Biophys. J.* **92**, 4233–4243.

Norton, R. S. and Pallaghy, P. K. (1998). The cystine knot structure of ion channel toxins and related polypeptides. *Toxicon* **36**, 1573–1583.

Odell, G. V., Fenton, A. W., Ownby, C. L., Doss, M. P. and Schmidt, J. O. (1999). The role of venom citrate. *Toxicon* **37**, 407–409.

Odell, G. V., Ferry, P. C., Vick, L. M., Fenton, A. W., Decker, L. S., Cowell, R. L., Ownby, C. L. and Gutierrez, J. M. (1998). Citrate inhibition of snake venom proteases. *Toxicon* **36**, 1801–1806.

Orivel, J., Redeker, V., Le Caer, J. P., Krier, F., Revol-Junelles, A. M., Longeon, A., Chaffotte, A., Dejean, A. and Rossier, J. (2001). Ponericins, new antibacterial and insecticidal peptides from the venom of the ant *Pachycondyla goeldii*. *J. Biol. Chem.* **276**, 17823–17829.

Oswald, R. E., Suchyna, T. M., McFeeters, R., Gottlieb, P. and Sachs, F. (2002). Solution structure of peptide toxins that block mechanosensitive ion channels. *J. Biol. Chem.* **277**, 34443–34450.

Palma, M. S., Itagaki, Y., Fujita, T., Hisada, M., Naoki, H. and Nakajima, T. (1997). Mass spectrometric structure determination of spider toxins: arginine containing acylpolyamines from venoms of Brazilian garden spider *Nephilengys cruentata*. *Nat. Toxins* **5**, 47–57.

Palma, M. S., Itagaki, Y., Fujita, T., Naoki, H. and Nakajima, T. (1998). Structural characterization of a new acylpolyaminetoxin from the venom of Brazilian garden spider *Nephilengys cruentata*. *Toxicon* **36**, 485–493.

Paludo, K. S., Biscaia, S. M., Chaim, O. M., Otuki, M. F., Naliwaiko, K., Dombrowski, P. A., Franco, C. R. and Veiga, S. S. (2009). Inflammatory events induced by brown spider venom and its recombinant dermonecrotic toxin: a pharmacological investigation. *Comp. Biochem. Physiol. C Toxicol. Pharmacol.* **149**, 323–333.

Pan, Z., Barry, R., Lipkin, A. and Soloviev, M. (2007). Selection strategy and the design of hybrid oligonucleotide primers for RACE-PCR: cloning a family of toxin-like sequences from *Agelena orientalis*. *BMC Mol. Biol.* **8**, 32.

Pansa, M. C., Natalizi, G. M. and Bettini, S. (1972). 5-hydroxytryptamine content of *Latrodectus mactans tredecimguttatus* venom from gland extracts. *Toxicon* **10**, 85–86.

Parsons, M. E. (1991). Histamine receptors: an overview. *Scand. J. Gastroenterol. Suppl.* **180**, 46–52.

Pekar, S., Toft, S., Hruskova, M. and Mayntz, D. (2007). Dietary and prey-capture adaptations by which *Zodarion germanicum*, an ant-eating spider (Araneae: Zodariidae), specialises on the Formicinae. *Naturwissenschaften* **95**, 233–239.

Peng, K., Lin, Y. and Liang, S.-P. (2006). Nuclear magnetic resonance studies on huwentoxin-XI from the Chinese bird spider *Ornithoctonus huwena*: ^{15}N labeling and Sequence-specific ^{1}H, ^{15}N nuclear magnetic resonance assignments. *Acta Biochim. Biophys. Sin.* **38**, 457–466.

Peng, K., Shu, Q., Liu, Z. and Liang, S. (2002). Function and solution structure of huwentoxin-IV, a potent neuronal tetrodotoxin (TTX)-sensitive sodium channel antagonist from Chinese bird spider *Selenocosmia huwena*. *J. Biol. Chem.* **277**, 47564–47571.

Penney, D. and Selden, P. A. (2011). Fossil Spiders: The Evolutionary History of a Mega-Diverse Order. Siri Scientific Press, Manchester.

Perret, B. A. (1977). Proteolytic activity of tarantula venoms due to contamination with saliva. *Toxicon* **15**, 505–510.

Picolo, G., Hisada, M., Moura, A. B., Machado, M. F., Sciani, J. M., Conceicao, I. M., Melo, R. L., Oliveira, V., Lima-Landman, M. T., Cury, Y., Konno, K. and Hayashi, M. A. (2010). Bradykinin-related peptides in the venom of the solitary wasp *Cyphononyx fulvognathus*. *Biochem. Pharmacol.* **79**, 478–486.

Piek, T. (1991). Neurotoxic kinins from wasp and ant venoms. *Toxicon* **29**, 139–149.

Pimenta, A. M. and De Lima, M. E. (2005). Small peptides, big world: biotechnological potential in neglected bioactive peptides from arthropod venoms. *J. Pept. Sci.* **11**, 670–676.

Pimenta, A. M., Rates, B., Bloch, C. Jr., Gomes, P. C., Santoro, M. M., de Lima, M. E., Richardson, M. and Cordeiro, M. N. (2005). Electrospray ionization quadrupole time-of-flight and matrix-assisted laser desorption/ionization tandem time-of-flight mass spectrometric analyses to solve micro-heterogeneity in post-translationally modified peptides from *Phoneutria nigriventer* (Aranea, Ctenidae) venom. *Rapid Commun. Mass Spectrom.* **19**, 31–37.

Pimentel, C., Choi, S. J., Chagot, B., Guette, C., Camadro, J. M. and Darbon, H. (2006). Solution structure of PcFK1, a spider peptide active against *Plasmodium falciparum*. *Protein Sci.* **15**, 628–634.

Piser, T. M., Lampe, R. A., Keith, R. A. and Thayer, S. A. (1995). Complete and reversible block by omega-grammotoxin SIA of glutamatergic synaptic transmission between cultured rat hippocampal neurons. *Neurosci. Lett.* **201**, 135–138.

Platnick, N. I. (2011). The World Spider Catalog, Version 11.5. Available at: http://research.amnh.org/entomology/spiders/catalog/index.html.

Powers, J. P. and Hancock, R. E. (2003). The relationship between peptide structure and antibacterial activity. *Peptides* **24**, 1681–1691.

Pukala, T. L., Boland, M. P., Gehman, J. D., Kuhn-Nentwig, L., Separovic, F. and Bowie, J. H. (2007a). Solution structure and interaction of cupiennin 1a, a spider venom peptide, with phospholipid bilayers. *Biochemistry* **46**, 3576–3585.

Pukala, T. L., Doyle, J. R., Llewellyn, L. E., Kuhn-Nentwig, L., Apponyi, M. A., Separovic, F. and Bowie, J. H. (2007b). Cupiennin 1a, an antimicrobial peptide from the venom of the neotropical wandering spider *Cupiennius salei*, also inhibits the formation of nitric oxide by neuronal nitric oxide synthase. *FEBS J.* **274**, 1778–1784.

Qadri, Y. J., Berdiev, B. K., Song, Y., Lippton, H. L., Fuller, C. M. and Benos, D. J. (2009). Psalmotoxin-1 docking to human acid-sensing ion channel-1. *J. Biol. Chem.* **284**, 17625–17633.

Rash, L. D. and Hodgson, W. C. (2002). Pharmacology and biochemistry of spider venoms. *Toxicon* **40**, 225–254.

Rash, L. D., King, R. G. and Hodgson, W. C. (1998). Evidence that histamine is the principal pharmacological component of venom from an Australian wolf spider (*Lycosa godeffroyi*). *Toxicon* **36**, 367–375.

Rash, L. D., King, R. G. and Hodgson, W. C. (2000). Sex differences in the pharmacological activity of venom from the white-tailed spider (*Lampona cylindrata*). *Toxicon* **38**, 1111–1127.

Redaelli, E., Cassulini, R. R., Silva, D. F., Clement, H., Schiavon, E., Zamudio, F. Z., Odell, G., Arcangeli, A., Clare, J. J., Alagon, A., de la Vega, R. C., Possani, L. D. and Wanke, E. (2010). Target promiscuity and heterogeneous effects of tarantula venom peptides affecting Na^+ and K^+ ion channels. *J. Biol. Chem.* **285**, 4130–4142.

Richardson, M., Pimenta, A. M., Bemquerer, M. P., Santoro, M. M., Beirão, P. S., Lima, M. E., Figueiredo, S. G., Bloch, C. Jr., Vasconcelos, E. A., Campos, F. A., Gomes, P. C. and Cordeiro, M. N. (2006). Comparison of the partial proteomes of the venoms of Brazilian spiders of the genus *Phoneutria*. *Comp. Biochem. Physiol. C Toxicol. Pharmacol.* **142**, 173–187.

Robinson, M. H. and Mirick, H. (1971). The predatory behavior of the golden-web spider *Nephila clavipes* (Araneae: Araneidae). *Psyche* **78**, 123–139.

Robinson, M. H., Mirick, H. and Turner, O. (1969). The predatory behavior of some araneid spiders and the origin of immobilization wrapping. *Psyche* **76**, 487–501.

Rocha-e-Silva, T. A. A., Sutti, R. and Hyslop, S. (2009). Milking and partial characterization of venom from the Brazilian spider *Vitalius dubius* (Theraphosidae). *Toxicon* **53**, 153–161.

Rodrigues, M. C., Guimaraes, L. H., Liberato, J. L., de Lourdes, Teixeira, de Moraes Polizeli, M. and dos Santos, W. F. (2006). Acid and alkaline phosphatase activities of a fraction isolated from *Parawixia bistriata* spider venom. *Toxicon* **47**, 854–858.

Rodrigues, M. C., Guizzo, R., Gobbo-Neto, L., Ward, R. J., Lopes, N. P. and dos Santos, W. F. (2004). The biological activity in mammals and insects of the nucleosidic fraction from the spider *Parawixia bistriata*. *Toxicon* **43**, 375–383.

Roeder, T. (1999). Octopamine in invertebrates. *Prog. Neurobiol.* **59**, 533–561.

Roeder, T. (2005). Tyramine and octopamine: ruling behavior and metabolism. *Annu. Rev. Entomol.* **50**, 447–477.

Rohou, A., Nield, J. and Ushkaryov, Y. A. (2007). Insecticidal toxins from black widow spider venom. *Toxicon* **49**, 531–549.

Rosengren, K. J., Wilson, D., Daly, N. L., Alewood, P. F. and Craik, D. J. (2002). Solution structures of the cis- and trans-Pro30 isomers of a novel 38-residue toxin from the venom of *Hadronyche infensa* sp. that contains a cystine-knot motif within its four disulfide bonds. *Biochemistry* **41**, 3294–3301.

Russell, F. E. (1966). Phosphodiesterase of some snake and arthropod venoms. *Toxicon* **4**, 153–154.

Saidemberg, D. M., Ferreira, M. A., Takahashi, T. N., Gomes, P. C., Cesar-Tognoli, L. M., da Silva-Filho, L. C., Tormena, C. F., da Silva, G. V. and Palma, M. S. (2009). Monoamine oxidase inhibitory activities of indolylalkaloid toxins from the venom of the colonial spider *Parawixia bistriata*: functional characterization of PwTX-I. *Toxicon* **54**, 717–724.

Salles, H. C., Volsi, E. C., Marques, M. R., Souza, B. M., dos Santos, L. D., Tormena, C. F., Mendes, M. A. and Palma, M. S. (2006). The venomous secrets of the web droplets from the viscid spiral of the orb-weaver spider *Nephila clavipes* (Araneae, Tetragnatidae). *Chem. Biodivers.* **3**, 727–741.

Sanguinetti, M. C., Johnson, J. H., Hammerland, L. G., Kelbaugh, P. R., Volkmann, R. A., Saccomano, N. A. and Mueller, A. L. (1997). Heteropodatoxins: peptides isolated from spider venom that block Kv4.2 potassium channels. *Mol. Pharmacol.* **51**, 491–498.

Santos, D., Verly, R., Piló-Veloso, D., de Maria, M., de Carvalho, M., Cisalpino, P., Soares, B., Diniz, C., Farias, L., Moreira, D., Frézard, F., Bemquerer, M., Pimenta, A. and de Lima, M. (2010). LyeTx I, a potent antimicrobial peptide from the venom of the spider *Lycosa erythrognatha*. *Amino Acids* **39**, 135–144.

Saraswati, S., Fox, L. E., Soll, D. R. and Wu, C. F. (2004). Tyramine and octopamine have opposite effects on the locomotion of *Drosophila* larvae. *J. Neurobiol.* **58**, 425–441.

Sasaki, S. D., Azzolini, S. S., Hirata, I. Y., Andreotti, R. and Tanaka, A. S. (2004). *Boophilus microplus* tick larvae, a rich source of Kunitz type serine proteinase inhibitors. *Biochimie* **86**, 643–649.

Sasaki, S. D. and Tanaka, A. S. (2008). rBmTI-6, a Kunitz-BPTI domain protease inhibitor from the tick *Boophilus microplus*, its cloning, expression and biochemical characterization. *Vet. Parasitol.* **155**, 133–141.

Satake, H., Villegas, E., Oshiro, N., Terada, K., Shinada, T. and Corzo, G. (2004). Rapid and efficient identification of cysteine-rich peptides by random screening of a venom gland cDNA library from the hexathelid spider *Macrothele gigas*. *Toxicon* **44**, 149–156.

Savel-Niemann, A. (1989). Tarantula (*Eurypelma californicum*) venom, a multicomponent system. *Biol. Chem. Hoppe Seyler* **370**, 485–498.

Savel-Niemann, A. and Roth, D. (1989). Biochemical analysis of tarantula venom (*Eurypelma californicum*). *Naturwissenschaften* **76**, 212–213.

Schäfer, A., Benz, H., Fiedler, W., Guggisberg, A., Bienz, S. and Hesse, M. (1994). Polyamine toxins from spiders and wasps. In: *The Alkaloids, Chemistry and Pharmacology* (eds Cordell, G. A. and Brossi, A.), pp. 1–125. Academic Press, San Diego.

Schanbacher, F. L., Lee, C. K. and Hall, J. E. (1973a). Composition and properties of tarantula *Dugesiella hentzi* (Girard) venom. *Toxicon* **11**, 21–29.

Schanbacher, F. L., Lee, C. K., Wilson, I. B., Howell, D. E. and Odell, G. V. (1973b). Purification and characterization of tarantula, *Dugesiella hentzi* (Girard) venom hyaluronidase. *Comp. Biochem. Physiol.* **44**, 389–396.

Scheiner, R., Baumann, A. and Blenau, W. (2006). Aminergic control and modulation of honeybee behaviour. *Curr. Neuropharmacol.* **4**, 259–276.

Schenberg, S. and Pereira Lima, F. A. (1978). Venoms of ctenids. In: *Arthropod Venoms. Handbook of Experimental Pharmacology* (ed Bettini, S.), pp. 217–244. Springer, Berlin.

Schenone, H., Saavedra, T., Rojas, A. and Villarroel, F. (1989). Loxoscelism in Chile. Epidemiological, clinical and experimental studies. *Rev. Inst. Med. Trop. Sao Paulo* **31**, 403–415.

Schenone, H. and Suarez, G. (1978). Venoms of Scytodidae. Genus *Loxosceles*. In: *Arthropod Venoms. Handbook of Experimental Pharmacology* (ed Bettini, S.), pp. 247–275. Springer, Berlin.

Schroeder, F. C., Taggi, A. E., Gronquist, M., Malik, R. U., Grant, J. B., Eisner, T. and Meinwald, J. (2008). NMR-spectroscopic screening of spider venom reveals sulfated nucleosides as major components for the brown recluse and related species. *Proc. Natl. Acad. Sci. USA* **105**, 14283–14287.

Schweitz, H., Bruhn, T., Guillemare, E., Moinier, D., Lancelin, J. M., Beress, L. and Lazdunski, M. (1995). Kalicludines and kaliseptine. Two different classes of sea anemone toxins for voltage sensitive K^+ channels. *J. Biol. Chem.* **270**, 25121–25126.

Shai, Y. (2002). Mode of action of membrane active antimicrobial peptides. *Biopolymers* **66**, 236–248.

Shanbrom, E. (2003). Use of citric acid as an antimicrobial agent or enhancer or as anticancer agent. (WIPO Patent Publ. Nr WO/2002/034293).

Sheumack, D. D., Baldo, B. A., Carroll, P. R., Hampson, F., Howden, M. E. H. and Skorulis, A. (1984). A comparative study of properties and toxic constituents of funnel web spider (*Atrax*) venoms. *Comp. Biochem. Physiol.* **78**, 55–68.

Shikata, Y., Watanabe, T., Teramoto, T., Inoue, A., Kawakami, Y., Nishizawa, Y., Katayama, K. and Kuwada, M. (1995). Isolation and characterization of a peptide isomerase from funnel web spider venom. *J. Biol. Chem.* **270**, 16719–16723.

Shu, Q. and Liang, S. P. (1999). Purification and characterization of huwentoxin-II, a neurotoxic peptide from the venom of the Chinese bird spider *Selenocosmia huwena*. *J. Pept. Res.* **53**, 486–491.

Shu, Q., Lu, S.-Y., Gu, X.-C. and Liang, S.-P. (2002). The structure of spider toxin huwentoxin-II with unique disulfide linkage: evidence for structural evolution. *Protein Sci.* **11**, 245–252.

Shultz, J. W. (1991). Evolution of locomotion in Arachnida: the hydraulic pressure pump of the giant whipscorpion, *Mastigoproctus giganteus* (Uropygi). *J. Morphol.* **210**, 13–31.

Siemens, J., Zhou, S., Piskorowski, R., Nikai, T., Lumpkin, E. A., Basbaum, A. I., King, D. and Julius, D. (2006). Spider toxins activate the capsaicin receptor to produce inflammatory pain. *Nature* **444**, 208–212.

Silva, P. I. Jr., Daffre, S. and Bulet, P. (2000). Isolation and characterization of gomesin, an 18-residue cysteine-rich defense peptide from the spider *Acanthoscurria gomesiana* hemocytes with sequence similarities to horseshoe crab antimicrobial peptides of the tachyplesin family. *J. Biol. Chem.* **275**, 33464–33470.

Skinner, W. S., Adams, M. E., Quistad, G. B., Kataoka, H., Cesarin, B. J., Enderlin, F. E. and Schooley, D. A. (1989). Purification and characterization of two classes of neurotoxins from the funnel web spider *Agelenopsis aperta*. *J. Biol. Chem.* **264**, 2150–2155.

Sosnina, N. A., Golubenko, Z., Akhunov, A. A., Kugaevskaia, E. V., Eliseeva Iu, E. and Orekhovich, V. N. (1990). Bradykinin-potentiating peptides from the spider *Latrodectus tredecimguttatus*– inhibitors of carboxycathepsin and of a preparation of karakurt venom kininase. *Dokl. Akad. Nauk SSSR* **315**, 236–239.

Stöcklin, R. and Cretton, G. (1998). *VENOMS – The Ultimate Database on Venomous Animals and Their Venoms*. Atheris Laboratories, Geneva, Switzerland (Professional edition, Version 1.0: CD-ROM 1 & 2).

Stöcklin, R. and Favreau, P. (2002). Proteomics of venom peptides. In: *Perspectives in Molecular Toxinology* (ed Ménez, A.), pp. 107–123. John Wiley & Sons Ltd, Chichester, UK.

Stöcklin, R. and Vorherr, T. (2010). Venoms – A natural source for mini-protein drugs. Pharmanufacturing: The International Peptide Review (Sept. 2010), pp. 44–46.

Strang, R. H., Whitton, P. S., Jabbar, A. and Nicholson, R. A. (1990). Taurine as a neuromodulator in the insect central nervous system. *Prog. Clin. Biol. Res.* **351**, 227–235.

Strydom, D. J. (1973). Protease inhibitors as snake venom toxins. *Nat. New Biol.* **243**, 88–89.

Suchyna, T. M., Johnson, J. H., Hamer, K., Leykam, J. F., Gage, D. A., Clemo, H. F., Baumgarten, C. M. and Sachs, F. (2000). Identification of a peptide toxin from *Grammostola spatulata* spider venom that blocks cation-selective stretch-activated channels. *J. Gen. Physiol.* **115**, 583–598.

Sutherland, S. K. (1978). Venoms of Dipluridae. In: *Arthropod Venoms. Handbook of Experimental Pharmacology* (ed Bettini, S.), pp. 121–148. Springer, Berlin.

Swartz, K. J. (2007). Tarantula toxins interacting with voltage sensors in potassium channels. *Toxicon* **49**, 213–230.

Taggi, A. E., Meinwald, J. and Schroeder, F. C. (2004). A new approach to natural products discovery exemplified by the identification of sulfated nucleosides in spider venom. *J. Am. Chem. Soc.* **126**, 10364–10369.

Tambourgi, D. V., Magnoli, F. C., Van den Berg, C. W., Morgan, B. P., De Araujo, P. S., Alves, E. W. and Da Silva, W. D. (1998). Sphingomyelinases in the venom of the spider *Loxosceles intermedia* are responsible for both dermonecrosis and complement-dependent hemolysis. *Biochem. Biophys. Res. Commun.* **251**, 366–373.

Tang, X., Zhang, Y., Hu, W., Xu, D., Tao, H., Yang, X., Li, Y., Jiang, L. and Liang, S. (2010). Molecular diversification of peptide toxins from the tarantula *Haplopelma hainanum* (*Ornithoctonus hainana*) venom based on transcriptomic, peptidomic, and genomic analyses. *J. Proteome Res.* **9**, 2550–2564.

Toki, T., Yasuhara, T., Aramaki, Y., Osawa, K., Miwa, A., Kawai, N. and Nakajima, T. (1988). Isolation and chemical characterization of a series of new spider toxin nephilatoxins in the venom of joro spider *Nephila clavata*. *Biomed. Res.* **9**, 421–428.

Torres-Larios, A., Gurrola, G. B., Zamudio, F. Z. and Possani, L. D. (2000). Hadrurin, a new antimicrobial peptide from the venom of the scorpion *Hadrurus aztecus*. *Eur. J. Biochem.* **267**, 5023–5031.

Tossi, A. and Sandri, L. (2002). Molecular diversity in gene-encoded, cationic antimicrobial polypeptides. *Curr. Pharm. Des.* **8**, 743–761.

Trachsel, C. (2010). Biochemical analysis and recombinant expression of venom components of the Central American hunting spider *Cupiennius salei*. PhD Thesis, University of Bern, Switzerland.

Trevisan-Silva, D., Gremski, L. H., Chaim, O. M., da Silveira, R. B., Meissner, G. O., Mangili, O. C., Barbaro, K. C., Gremski, W., Veiga, S. S. and Senff-Ribeiro, A. (2010). Astacin-like metalloproteases are a gene family of toxins present in the venom of different species of the brown spider (genus *Loxosceles*). *Biochimie* **92**, 21–32.

Tzouros, M., Chesnov, S., Bienz, S., Hesse, M. and Bigler, L. (2005). New linear polyamine derivatives in spider venoms. *Toxicon* **46**, 350–354.

Usherwood, P. N. R. (1994). Targeting locust muscle glutamate receptors with polyamine-containing toxins. In: *Natural and Engineered Pest Managment Agents* (eds Hedin, P. A., Menn, J. J. and Hollingworth, R. M.), pp. 259–267. American Chemical Society.

Usherwood, P. N. R. and Blagbrough, I. S. (1989). Antagonsim of insect muscle glutamate receptors with particular reference to arthropod toxins. In: *Insecticide Action, From Molecule to Organism* (eds Narahashi, T. and Chambers, J.), pp. 13–31. Plenum Press, New York.

Usherwood, P. N. R., Duce, I. R. and Boden, P. (1984). Slowly-reversible block of glutamate receptor-channels by venoms of the spiders, *Argiope trifasciata* and *Araneus gemma*. *J. Physiol.* **79**, 241–245.

Usmanov, P. B. and Nuritova, F. A. (1994). The anticoagulant action of phospholipase A from *Eresus niger* spider venom. *Toxicon* **32**, 625–628.

Vassilevski, A. A., Fedorova, I. M., Maleeva, E. E., Korolkova, Y. V., Efimova, S. S., Samsonova, O. V., Schagina, L. V., Feofanov, A. V., Magazanik, L. G. and Grishin, E. V. (2010). Novel class of spider toxin: active principle from the yellow sac spider *Cheiracanthium punctorium* venom is a unique two-domain polypeptide. *J. Biol. Chem.* **285**, 32293–32302.

Vassilevski, A. A., Kozlov, S. A. and Grishin, E. V. (2009). Molecular diversity of spider venom. *Biochemistry (Moscow)* **74**, 1505–1534.

Vassilevski, A. A., Kozlov, S. A., Samsonova, O. V., Egorova, N. S., Karpunin, D. V., Pluzhnikov, K. A., Feofanov, A. V. and Grishin, E. V. (2008). Cyto-insectotoxins, a novel class of cytolytic and insecticidal peptides from spider venom. *Biochem. J.* **411**, 687–696.

Veiga, S. S., da Silveira, R. B., Dreyfus, J. L., Haoach, J., Pereira, A. M., Mangili, O. C. and Gremski, W. (2000). Identification of high molecular weight serine-proteases in *Loxosceles intermedia* (brown spider) venom. *Toxicon* **38**, 825–839.

Vemana, S., Pandey, S. and Larsson, H. P. (2008). Intracellular Mg^{2+} is a voltage-dependent pore blocker of HCN channels. *Am. J. Physiol. Cell Physiol.* **295**, C557–C565.

Venema, V. J., Swiderek, K. M., Lee, T. D., Hathaway, G. M. and Adams, M. E. (1992). Antagonism of synaptosomal calcium channels by subtypes of ω-Agatoxins. *J. Biol. Chem.* **267**, 2610–2615.

Verano-Braga, T., Rocha-Resende, C., Silva, D. M., Ianzer, D., Martin-Eauclaire, M. F., Bougis, P. E., de Lima, M. E., Santos, R. A. and Pimenta, A. M. (2008). *Tityus serrulatus* hypotensins: a new family of peptides from scorpion venom. *Biochem. Biophys. Res. Commun.* **371**, 515–520.

Villegas, E., Adachi-Akahane, S., Bosmans, F., Tytgat, J., Nakajima, T. and Corzo, G. (2008). Biochemical characterization of cysteine-rich peptides from *Oxyopes* sp. venom that block calcium ion channels. *Toxicon* **52**, 228–236.

Vollrath, F., Fairbrother, W. J., Williams, R. J. P., Tillinghast, E. K., Bernstein, D. T., Gallager, K. S. and Townley, M. A. (1990). Compounds in the droplets of the orb spider's viscid spiral. *Nature* **345**, 526–528.

Volsi, E. C., Mendes, M. A., Marques, M. R., dos Santos, L. D., Santos, K. S., de Souza, B. M., Babieri, E. F. and Palma, M. S. (2006). Multiple bradykinin-related peptides from the capture web of the spider *Nephila clavipes* (Araneae, Tetragnatidae). *Peptides* **27**, 690–697.

Wang, X., Connor, M., Smith, R., Maciejewski, M. W., Howden, M. E., Nicholson, G. M., Christie, M. J. and King, G. F. (2000). Discovery and characterization of a family of insecticidal neurotoxins with a rare vicinal disulfide bridge. *Nat. Struct. Biol.* **7**, 505–513.

Wang, Z. and Wang, G. (2004). APD: the antimicrobial peptide database. *Nucleic Acids Res.* **32**, D590–D592.

Welsh, J. H. and Batty, C. S. (1963). 5-Hydroxytryptamine content of some arthropod venoms and venom-containing parts. *Toxicon* **1**, 165–173.

Weng, J. L., Barrantes, G. and Eberhard, W. G. (2006). Feeding by *Philoponella vicina* (Araneae, Uloboridae) and how uloborid spiders lost their venom glands. *Can. J. Zool.* **84**, 1752–1762.

Wigger, E., Kuhn-Nentwig, L. and Nentwig, W. (2002). The venom optimisation hypothesis: a spider injects large venom quantities only into difficult prey types. *Toxicon* **40**, 749–752.

Williams, K. (1997a). Interactions of polyamines with ion channels. *Biochem. J.* **325**, 289–297.

Williams, K. (1997b). Modulation and block of ion channels: a new biology of polyamines. *Cell. Signal.* **9**, 1–13.

Witte, I., Kreienkamp, H. J., Gewecke, M. and Roeder, T. (2002). Putative histamine-gated chloride channel subunits of the insect visual system and thoracic ganglion. *J. Neurochem.* **83**, 504–514.

Woolfson, D. N. and Williams, D. H. (1990). The influence of proline residues on α-helical structure. *FEBS Lett.* **277**, 185–188.

Wright, R. P., Elgert, K. D., Campbell, B. J. and Barrett, J. T. (1973). Hyaluronidase and esterase activities of the venom of the poisonous brown recluse spider. *Arch. Biochem. Biophys.* **159**, 415–426.

Wullschleger, B., Kuhn-Nentwig, L., Tromp, J., Kämpfer, U., Schaller, J., Schürch, S. and Nentwig, W. (2004). CSTX-13, a highly synergistically acting two-chain

neurotoxic enhancer in the venom of the spider *Cupiennius salei* (Ctenidae). *Proc. Natl. Acad. Sci. USA* **101**, 11251–11256.
Wullschleger, B., Nentwig, W. and Kuhn-Nentwig, L. (2005). Spider venom: enhancement of venom efficacy mediated by different synergistic strategies in *Cupiennius salei. J. Exp. Biol.* **208**, 2115–2121.
Yan, L. Z. and Adams, M. E. (1998). Lycotoxins, antimicrobial peptides from venom of the wolf spider *Lycosa carolinensis. J. Biol. Chem.* **273**, 2059–2066.
Young, A. R. and Pincus, S. J. (2001). Comparison of enzymatic activity from three species of necrotising arachnids in Australia: *Loxosceles rufescens, Badumna insignis* and *Lampona cylindrata. Toxicon* **39**, 391–400.
Yuan, C., Jin, Q., Tang, X., Hu, W., Cao, R., Yang, S., Xiong, J., Xie, C., Xie, J. and Liang, S. (2007a). Proteomic and peptidomic characterization of the venom from the Chinese bird spider, *Ornithoctonus huwena* Wang. *J. Proteome Res.* **6**, 2792–2801.
Yuan, C., Liao, Z., Zeng, X., Dai, L., Kuang, F. and Liang, S. (2007b). Jingzhaotoxin-XII, a gating modifier specific for Kv4.1 channels. *Toxicon* **50**, 646–652.
Yuan, C. H., He, Q. Y., Peng, K., Diao, J. B., Jiang, L. P., Tang, X. and Liang, S. P. (2008). Discovery of a distinct superfamily of Kunitz-type toxin (KTT) from tarantulas. *PLoS One* **3**, e3414.
Yusein, S., Wolstenholme, A. and Semenov, E. (2010). Functional consequences of mutations in the *Drosophila* histamine receptor HCLB. *J. Insect Physiol.* **56**, 21–27.
Zhang, Y., Chen, J., Tang, X., Wang, F., Jiang, L., Xiong, X., Wang, M., Rong, M., Liu, Z. and Liang, S. (2010). Transcriptome analysis of the venom glands of the Chinese wolf spider *Lycosa singoriensis. Zoology (Jena)* **113**, 10–18.
Zhu, S., Darbon, H., Dyason, K., Verdonck, F. and Tytgat, J. (2003). Evolutionary origin of inhibitor cystine knot peptides. *FASEB J.* **17**, 1765–1767.
Zonstein, S. L. and Ovtchinnikov, S. V. (1999). A new Central Asian species of the spider genus *Lachesana* Strand, 1932 (Araneae, Zodariidae: Lachesaninae). *TETHYS Entomol. Res.* **1**, 59–62.

Spider Nutrition: An Integrative Perspective

Shawn M. Wilder

School of Biological Sciences, The University of Sydney, Sydney, New South Wales, Australia

Abstract

Spiders represent a diverse, widespread and abundant group of carnivores. Studying the nutritional ecology of spiders is critical because it can aid in understanding the evolution of prey capture and life history strategies, factors regulating the abundance and diversity of spiders in particular habitats and the role of spiders in arthropod community dynamics including biological control of crop pests. The feeding habits of spiders have long attracted the attention of biologists, in part, because many build webs and are relatively easy to observe. While these studies have provided a wealth of information on the

ADVANCES IN INSECT PHYSIOLOGY VOL. 40
ISBN 978-0-12-387668-3
DOI: 10.1016/B978-0-12-387668-3.00002-7

abundance and types of prey captured, they have yielded relatively little information on the types and quantities of nutrients ingested by spiders in nature. Relatively little is also known about the nutritional requirements of spiders, although recent studies using more controlled manipulations of prey nutrient content are beginning to provide a clearer understanding of the types and quantities of nutrients needed by spiders to maximize performance. There is a tremendous opportunity to rapidly advance our understanding of spider nutrition, given a strong foundation in the natural history, behaviour, physiology and ecology of spiders and recent advances in analytical techniques and frameworks for studying nutrition. Studies focusing on the connections between spider nutritional physiology and how this is affected by and affects prey communities may produce particularly exciting results.

1 Introduction

Food resources are essential for all aspects of an animal's life and the amount and types of food consumed can have large influences on growth, reproduction and survival (Raubenheimer and Simpson, 1997; Raubenheimer *et al.*, 2009; Slansky and Rodriguez, 1987; Sterner and Elser, 2002; Waldbauer, 1968; White, 1978). The field of nutritional ecology encompasses the study of the dynamic interactions between food composition and animal physiology, ecology and evolution (Raubenheimer *et al.*, 2009). While this field has been around for decades (Slansky and Rodriguez, 1987; Waldbauer, 1968), there has recently been a resurgence of interest in nutrition and the development of rigorous theoretical and quantitative frameworks to study nutrition (Raubenheimer *et al.*, 2009; Sterner and Elser, 2002). Recent studies have revealed exciting findings such as connections between food composition, mRNA production and growth of arthropods (Elser *et al.*, 2000, 2006; Sterner and Elser, 2002), metabolism, herbivore diet selection and plant community composition (Hawlena and Schmitz, 2010; Schmitz, 2003), and even food prices, protein content and human obesity (Brooks *et al.*, 2010; Simpson and Raubenheimer, 2005, 2009). However, much of the research on nutritional ecology has focused on herbivores and omnivores and far less research has been conducted on carnivores. A more complete understanding of carnivore nutrition and how it impacts trophic interactions is critical for building models of community and ecosystem dynamics and understanding the ecological and evolutionary consequences of changes in trophic interactions in communities due to extinctions, invasive species and climate change (Simpson *et al.*, 2009).

1.1 OBJECTIVES

There have been several previous reviews of spider nutrition, which have provided a strong foundation in the digestive physiology and foraging behaviour of spiders (Foelix, 1996; Nentwig, 1987; Riechert and Harp, 1987). However, important recent advances in the study of nutrition have introduced rigorous experimental frameworks that allow for greater integration between macronutrients in food, animal physiology and performance (Raubenheimer *et al.*, 2009). In addition, recent advances in methods to manipulate specific nutrients for spiders now allow for the opportunity to conduct more precise and rigorous experiments on spider nutrition (Jensen *et al.*, 2011a,b; Mayntz *et al.*, 2005). There is a tremendous opportunity to use these relatively new frameworks and techniques to rapidly advance our understanding of which nutrients are required in the diet of spiders, how spiders obtain those nutrients in nature and the consequences of nutritional deficiencies for spider performance and diet selection, which can ultimately impact arthropod community composition and ecosystem function (Table 1).

The goals of this synthesis are to review what is known about spider nutrition, highlight recent advances and how they can help provide a more integrative and in-depth understanding of spider nutrition and describe frameworks that may help to guide future research (Tables 1 and 2). While this synthesis is focused on spiders, many of these ideas may also be applicable to carnivore nutrition in general. Relatively little is known about spider nutrition, especially compared to other arthropods like herbivores and omnivores (e.g. Behmer, 2009; Lee *et al.*, 2008; Maklakov *et al.*, 2008). Hence, at times, this synthesis will be speculative and propose hypotheses that have little direct support from spiders but may be based off of ideas from related animals (e.g. other arthropods or other carnivores) or long-held beliefs about differences between spiders and other animals. These hypotheses and speculation are not meant to be conclusive statements about spider biology but rather as starting points or *a priori* expectations that need to be rigorously tested in future studies (Table 1).

1.2 WHY SPIDERS?

Spiders (Arachnida, Araneae) represent one of the largest groups of obligate carnivores with almost 40,000 recorded species (Ubick *et al.*, 2005) and are found in nearly all terrestrial communities (Foelix, 1996; Riechert, 1974). In addition, spiders can have a large influence on the structure and function of ecological communities through their direct and indirect effects on herbivores and omnivores (Fagan *et al.*, 2002; Hurd and Eisenberg, 1990; Krivan and Schmitz, 2004; Moran and Hurd, 1998; Riechert, 1974; Schmitz, 1998; Schmitz, 2004; Van Hook, 1971). Even just the presence or chemical cues of a spider can cause herbivores to shift their feeding behaviour to reduce overall consumption of plants or change the types of plants consumed, with resulting effects on plant community composition and nutrient cycling (Hawlena and Schmitz, 2010; Schmitz, 2003). In addition to natural systems, spiders may also have important effects on lower trophic levels in agricultural ecosystems (Riechert and Lockley, 1984). Increased densities of spiders in agroecosystems can result in herbivore suppression (Schmitz *et al.*, 2000), reduced herbivory (Hlivko and Rypstra, 2003) and increased crop yield (Riechert and Bishop,

TABLE 1
Several important questions that need to be addressed in the study of spider nutrition

Key questions in spider nutrition	Relevant citations
Important aspects of spider physiology	
Why do spiders have high nitrogen content in their bodies?	Wise (1993), Fagan *et al.* (2002), Wilder and Eubanks (2010)
What scale of nutrients is relevant for spiders?	Greenstone (1979), Higgins and Rankin (1999), Jensen *et al.* (2011a,b)
How important are carbohydrates or other plant-based food resources for spiders?	Jackson *et al.* (2001), Meehan *et al.* (2009), Taylor and Pfannenstiel (2009), Chen *et al.* (2010)
How does the nutrition of spiders compare to other carnivores?	MacDonald *et al.* (1984), Zoran (2002), Mayntz *et al.* (2009)
Do spiders store nitrogen or protein in their body, and if so, what form do they use?	Nentwig (1987), Tefler and Kunkel (1991)
The food of spiders in nature	
Are spiders food limited, nutrient limited or both in nature?	Wise (1993), Fagan *et al.* (2002), Raubenheimer *et al.* (2009), Wilder and Eubanks (2010)
What factors predict the nutrient content of insect prey?	Fagan *et al.* (2002), Lease and Wolf (2011)
Is there a simple way to estimate what nutrients in an insect body are available for spiders to ingest without testing every insect?	Wilder *et al.* (2010)
Can spiders select prey based on their nutritional composition?	Greenstone (1979), Toft (1999), Mayntz *et al.* (2005)
Nutrition and spider performance	
What intraspecific factors (e.g. sex, developmental stage) affect the nutritional requirements of spiders?	Uhl *et al.* (2004), Raubenheimer *et al.* (2007)
What interspecific factors (e.g. type of web built, activity level) affect the nutritional requirements of spiders?	Jensen *et al.* (2011a)
Is there genetic variation in the response of spiders to nutrients?	Uhl *et al.* (2004)
What is the chemical identity of prey toxins that affect spiders and how frequently are they found in prey?	Toft and Nielsen (1997), Fisker and Toft (2004)
How does the nutrient content of manipulated prey in laboratory experiments compare to prey in the field?	Jensen *et al.* (2011a,b)
General	
What are the consequences of spider nutrition for populations and communities of insects?	
Can spider nutrition inform our understanding of the role of spiders in biological control?	

TABLE 2
Summary of studies examining the effect of spider diet on performance

Family	Species	Performance measure	Exp.	Effect	Citation
Prey species					
Linyphiidae	*Erigone atra*	Consumption and fecundity	1	Differences among prey treatments	Marcussen *et al*. (1999)
Linyphiidae	*Atypena formosana*	Growth, survival and fecundity	1	Benefit of diet mixing	Sigsgaard *et al*. (2001)
Lycosidae	*Schizocosa* sp.	Growth and survival	1	Differences among prey treatments, some prey species were toxic	Toft and Wise (1999)
Lycosidae	*Pardosa prativaga*	Consumption, growth and survival	1	Differences among prey treatments, some prey species were toxic	Fisker and Toft (2004)
Lycosidae	*Pardosa lugubris*	Fecundity	1	No benefit of mixing, differences among prey species, some prey were toxic	Rickers *et al*. (2006)
Lycosidae	*Pardosa lugubris*	Growth and survival	1	Variable effects of mixed diet, some prey were toxic	Oelbermann and Scheu (2002a)
Lycosidae	*Pardosa lugubris*	Stable Isotope enrichment	1	Isotope enrichment differed depending upon the prey consumed	Oelbermann and Scheu (2002b)
Lycosidae	*Hogna helluo*	Growth and survival	1	Benefit of diet mixing	Uetz *et al*. (1992)
Lycosidae	*Hogna* sp.	Growth and survival	1	Benefit of diet mixing	Uetz *et al*. (1992)
Sicariidae	*Loxosceles laeta*	Growth and survival	1	Greater growth and survival on wide variety of prey	Lowrie (1987)
Lycosidae	*Pardosa prativaga*	Growth and survival	1	Despite early nutritional stress period on experimental diets, all spiders reached a similar final size	Jespersen and Toft (2003)

(*continues*)

TABLE 2 (*Continued*)

Family	Species	Performance measure	Exp.	Effect	Citation
Salticidae	*Portia fimbriata*	Growth and survival	1	Spiders performed best on diet of spiders only	Li and Jackson (1997)
Lycosidae	*Pardosa prativaga*	Enzyme activity	1	Spiders had higher detoxifying enzyme activity when their diet included toxic prey	Nielsen and Toft (2000)
Zodariidae	*Zodarion rubidum*	Growth and survival	2	Spiders grew larger and survived better when only feeding on ant foreparts	Pekár *et al.* (2010)
Salticidae	*Phidippus audax*	Consumption and growth	1	Spiders consumed more and gained more weight when feeding on mealworms	Strohmeyer *et al.* (1998)
Zodariidae	*Zodarion atlanticum* and *Z. germanicum*	Growth and survival	1	These ant-specialist spiders grew larger and survived better on the ant diet than the fly diet	Pekár and Toft (2009)
General nutrient manipulations					
Lycosidae	*Pardosa milvina*	Female aggression	3	Higher aggression when fed flies raised on dogfood-supplemented media	Wilder and Rypstra (2008b)
Lycosidae	*Pardosa prativaga*	Consumption	3	Greater consumption of a poor-quality prey after feeding on a high-quality diet	Mayntz and Toft (2000)
Lycosidae	*Pardosa amentata*	Growth and survival	4	Adding amino acids, fatty acids and dogfood to diet of prey increased spider growth	Mayntz and Toft (2001)

Lycosidae	*Pardosa prativaga*	Courtship, mating and male competition	3	Males on a high-quality diet had greater courtship intensity and higher mating success	Lomborg and Toft (2009)
Araneidae	*Zygiella x-notata*	Growth	3	Interactive effects of prey quality, quantity and predator sex	Mayntz *et al.* (2003)
Lycosidae	*Pardosa prativaga*	Growth, survival and cannibalism	3	Spiders grew larger and were more cannibalistic on the high-quality diet	Mayntz and Toft (2006)
Lycosidae	*Pardosa lugubris*	Body composition	5	Fatty acid composition of spiders reflected what was present in their prey	Pollierer *et al.* (2010)
Araneidae	*Araneus diadematus*	Survival and web production	6	Spiderlings survived longer and more frequently spun webs when supplemented with pollen	Smith and Mommsen (1984)
Miturgidae	*Cheiracanthium inclusum*	Growth, survival and fecundity	7	Spiders grew better, survived longer and produced more eggs when provided with nectar	Taylor and Pfannenstiel (2009)
Lycosidae	*Pardosa prativaga*	Survival and acetylcholinesterase activity	3	Synergistic effects of starvation and prey nutrients on response of spiders to insecticide	Pedersen *et al.* (2002)
Thomisidae	*Thomisus onustus*	Survival	8	Pollen and sucrose solutions aided survival	Vogelei and Greissl (1989)
Prey nutrients					
Lycosidae	*Pardosa prativaga*	Growth, metabolic rate and body composition	9	Greater growth on high-protein diet and longer survival on high-lipid diet	Jensen *et al.* (2010)

(*continues*)

TABLE 2 (*Continued*)

Family	Species	Performance measure	Exp.	Effect	Citation
Lycosidae	*Pardosa prativaga*	Consumption	9	Greater consumption of prey with opposite nutrients as past diet	Mayntz *et al.* (2005)
Eresidae	*Stegodyphus lineatus*	Consumption	9	Selective extraction of nitrogen from prey when previously fed low-protein diets	Mayntz *et al.* (2005)
Erisidae	*Stegodyphus lineatus*	Growth and consumption	10	Lipid-supplemented colonies of spiders had more adult females and females that were larger; adult females extracted more lipid from prey than juveniles	Salomon *et al.* (2008)
Thomisidae	*Misumenoides formosipes*	Consumption and survival	11	Male spiders preferred water containing sucrose and survived longer when given water with sucrose compared to only water	Pollard *et al.* (1995)
Miturgidae	*Cheiracanthium mildei*	Growth, survival and activity	11	Spiders were more likely to moult, had higher survival and were more active when provided nectar and sugar compared to water	Taylor and Bradley, (2009
Anyphaenidae	*Hibana velox*	Growth, survival and activity	11	Spiders were more likely to moult, had higher survival and were more active when provided nectar and sugar compared to water	Taylor and Bradley (2009)
Salticidae	*Portia quei*	Consumption, growth and body composition	9	Spiders fed high-protein prey had similar mass gain but consumed more food and had less lipid in their bodies	Toft *et al.* (2010)

Thomisidae	*Thomisus onustus*	Survival	11	Spiders fed sucrose had higher survival	Vogelei and Greissl (1989)
Araneidae	*Argiope keyserlingi*	Body condition and web structure	9	Spiders fed high-energy flies had better body condition but produced shorter web decorations	Blamires *et al.* (2009)
Araneidae	*Argiope keyserlingi*	Fecundity	9	Spiders fed high-energy flies had higher energy density in their eggs, there were no other effects of fly types of fecundity	Blamires (2010)
Lycosidae	*Pardosa amentata*	Growth, growth rate and instar duration	9	Spiders fed high-protein prey had higher growth rate	Jensen *et al.* (2011a,b)
Tetragnathidae	*Nephila clavipes*	Web properties	12	Less choline in web when there is less in the diet	Higgins and Rankin (1999)

Experimental manipulations: 1, fed a range of different prey species, including single and/or mixed species treatments; 2, different body parts of ants—the whole ant, just the foreparts or just the gaster; 3, fed *Drosophila melanogaster* that had been raised on nutrient-poor or dogfood-supplemented media; 4, fed *Drosophila melanogaster* that had been raised on nutrient-poor media supplemented with various nutrients including amino acids, fatty acids and vitamins; 5, fed Collembola that had been raised on different diets (fungus, leaves, Gram-negative and Gram-positive bacteria); 6, pollen or fungal spores were added to the spider web to be consumed when the web was recycled; 7, diet of lepidopteran eggs with or without nectar; 8, fed pollen, sucrose or *Drosophila melanogaster*; 9, fed flies (either *Drosophila melanogaster* or *M. domestica*) that had been raised on nutrient-poor media supplemented with various concentrations of carbohydrates (sucrose) or protein (casein); 10, fed crickets that had been raised on diets rich in protein (casein) or lipid (butter); 11, fed liquid diets containing carbohydrates (sucrose); 12, fed liquid diets of glucose and vitellin either with or without glycine and choline.

1990; Snyder and Wise, 2001). The effects of spiders on natural and agricultural ecosystems depend upon the abundance, habitat use and foraging activity of spiders (e.g. Riechert and Bishop, 1990; Schmitz, 1998; Settle *et al.*, 1996), and hence, it is important to understand the factors (e.g. nutrition) that influence spider foraging, growth and reproduction.

Interestingly, while the effects of spiders on community composition and ecosystem functioning are due to the feeding behaviour of spiders, relatively little is known about the nutritional ecology of spiders (e.g. what nutrients are required for optimal growth, reproduction and survival and do spiders obtain the required nutrients in nature?). The relative lack of research on spider nutrition could be due in part to the widespread beliefs that spiders are polyphagous and indiscriminate predators and that the nutritional composition of food is relatively unimportant for predators because all prey are of relatively high quality, at least compared to herbivores feeding on plants (Riechert and Harp, 1987). Yet recent studies are challenging these long-held assumptions about the nutritional ecology of spiders and carnivores, in general (Jensen *et al.*, 2011a,b; Mayntz *et al.*, 2005; Raubenheimer *et al.*, 2007; Wilder and Eubanks, 2010). Insect prey can vary in nutritional content and nutrients in prey can have a large effect on spider performance (Banjo *et al.*, 2006; Mayntz and Toft, 2001; Wilder *et al.*, 2010). Dynamic feedbacks between spider nutrition and prey nutrient content have the potential to cascade through food webs from the top down (e.g. changes in cannibalism, intraguild predation and direct and indirect effects on herbivore and plant communities; Elgar and Crespi, 1992; Hawlena and Schmitz, 2010; Polis and Holt, 1992; Polis *et al.*, 1989) or the bottom up (e.g. changes in the quantity, nutrient composition and distribution of prey carcasses and spider excreta; Schmitz *et al.*, 2010).

In addition to being interesting in their own right, spiders represent a model system for studying carnivore nutrition in general. There are few groups of carnivorous animals as diverse and widespread as spiders and much is known about the natural history, physiology, ecology and evolution of many spiders (Barth, 1985; Foelix, 1996; Nentwig, 1987; Ubick *et al.*, 2005; Wise, 1993). Spiders are also ideal candidates for laboratory and field research due to their sit-and-wait foraging habits, which means that they require much less space and change locations less frequently than other carnivore groups (Foelix, 1996; Uetz, 1992). In terms of evolutionary research, there is wide variation among spiders species in many traits including levels of activity (e.g. wandering vs. web building Blackledge *et al.*, 2009), sexual size dimorphism (Head, 1995; Hormiga *et al.*, 2000), reproductive strategies (Fromhage *et al.*, 2005; Miller, 2007), diet specialization (Li and Jackson, 1996; Meehan *et al.*, 2009) and mimicry (Jackson, 1992; Nelson and Jackson, 2006). Many of these traits have evolved independently multiple times, which facilitates the use of rigorous evolutionary and comparative analyses (e.g. Grafen, 1989; Purvis and Rambaut, 1995). This combination of characteristics (e.g. natural history information, ease of use in experiments, extraoral digestion and evolutionarily diverse traits) facilitates integrative studies of spider nutrition that connect physiology, ecology and evolution in both the laboratory and the field.

2 Important aspects of spider physiology

In this section, I focus on several aspects of spider physiology that may be important to understanding and studying spider nutrition. First, I provide a brief general overview of the importance of carnivory for spider nutrition and digestion of food by spiders. Then I discuss several aspects of spider physiology that may affect the nutritional requirements or allocation of nutrients by spiders compared to other animals, including nutrient storage, use of silk and metabolic rate.

2.1 CARNIVORY

Most studies of animal nutrition have been conducted on herbivores and omnivores. There are several important distinctions that need to be made when studying obligate carnivores such as spiders. The first is that protein is the most abundant macronutrient in the diet of carnivores, in contrast to herbivores in which carbohydrates may be far more abundant than protein. The high protein, and relatively low lipid and carbohydrate, content of spider diets means that spiders, like other carnivores, likely rely on catabolism of amino acids and gluconeogenesis to supply a significant proportion of their energy (Eisert, 2011). In herbivorous and omnivorous arthropods, high-protein diets can have deleterious consequences, especially for longevity or survival (Lee *et al.*, 2008; Maklakov *et al.*, 2008; Simpson and Raubenheimer, 2009). However, the long evolutionary history of carnivores feeding on high-protein diets has likely selected for adaptations that allow them to metabolize high-protein diets with limited consequences, which is an intriguing hypothesis that needs to be investigated further. The strong ability of spiders to catabolize amino acids for energy also likely provides them with some flexibility in their diet and the ability to overingest high-protein prey to gain limited energy (Jensen *et al.*, 2011a,b). However, the ability to catabolize amino acids for energy does not preclude the importance of lipids as a source of energy, especially given their high energy density (Horowitz, 2002).

In herbivores and omnivores, carbohydrates can be an important source of energy (Behmer, 2009). However, the role of carbohydrates in the nutrition of carnivores is relatively unclear. The digestion and utilization of carbohydrates by cats has been studied more intensively than any other carnivore (Buddington *et al.*, 1991; Eisert, 2011; Hewson-Hughes *et al.*, 2011). Cats tightly regulate the amount of carbohydrate in their diet below a certain level, even if this results in overall deficiencies in caloric intake (Hewson-Hughes *et al.*, 2011). This is likely because cats have a reduced ability to digest carbohydrates relative to other mammals and can suffer from obesity, decreased insulin sensitivity and acidosis of the bowel when feeding on high-carbohydrate diets (Hewson-Hughes *et al.*,

2011; Hoenig *et al.*, 2000). While much less is known about carbohydrate metabolism in spiders, some species appear to be able to digest and benefit from significant amounts of carbohydrates. In particular, recent work has shown that some spiders, especially highly active wandering spiders, seek out and ingest carbohydrates and that these carbohydrate resources can increase spider growth, reproduction and survival (Chen *et al.*, 2010; Jackson *et al.*, 2001; Meehan *et al.*, 2009; Pollard *et al.*, 1995; Taylor and Bradley, 2009; Vogelei and Greissl, 1989). In addition to those spiders that directly ingest carbohydrates from plants (e.g. from extrafloral nectaries), many more species could consume significant amounts of carbohydrates by consuming insect prey that feed on large quantities of carbohydrates such as some Diptera, Hymenoptera and Lepidoptera. Measurements and manipulations of carbohydrate concentrations in prey are needed to test the importance of this macronutrient for spider nutrition.

Carnivores could also differ from herbivores in the presence of certain biosynthetic pathways for dietary essential nutrients. For example, cats either lack enzymatic pathways or have a reduced capacity to synthesize a number of compounds (e.g. taurine, arginine, methionine and cysteine) compared to herbivorous mammals (Eisert, 2011; MacDonald *et al.*, 1984; Zoran, 2002). The evolutionary loss or reduction in biosynthetic pathways in carnivores is thought to be due to differences in the presence of these compounds in animal tissue compared to plants. If herbivores biosynthesize compounds and incorporate them into their bodies, then there may be less of a need for carnivores to biosynthesize these same compounds because they will be present in prey. This suggests that carnivores could have more dietary essential nutrients (e.g. amino acids, fatty acids) than herbivores. Unfortunately, very little is known about the role of dietary essential nutrients or specific amino or fatty acids for the nutrition of spiders despite the tremendous potential for these nutrients to affect spider performance or prey selection (Craig *et al.*, 2000; Grandison *et al.*, 2009; Greenstone, 1979; Pollierer *et al.*, 2010; Stanley-Samuelson *et al.*, 1988).

2.2 DIGESTION

Spiders feed using extraoral digestion (Cohen, 1995). This process involves injecting enzymes into prey, waiting for the enzymes to liquefy nutrients and then sucking nutrients out of the prey (Cohen, 1995). In the wolf spider *Pardosa prativaga*, the increase in metabolic rate during feeding (e.g. a 20-h period) accounts for 21% of the energetic content of a prey item, which includes the energetic costs of consuming, digesting and absorbing nutrients from the prey (Jensen *et al.*, 2010). Extraoral digestion effectively separates edible nutrients from inedible parts of insects and allows spiders to maximize their nutrient intake and minimize the ingestion of inedible bulk when they capture large prey. Spiders even have setae and other fine structures in their mouthparts that filter out very small particulate matter during feeding (Foelix, 1996). Extraoral digestion is different from other types of feeding in that it involves a large initial investment by the spider (i.e. injection of protein-rich enzymes). Given this investment, there may be selection for spiders to be choosy of the prey that they capture.

Extraoral digestion by spiders usually takes one of two different forms. In some species, including many web-building spiders, individuals will suck out the liquefied nutrients from the prey with little damage to the exoskeleton. In other cases, spiders will masticate the prey item into a ''meatball'' from which they will extract nutrients (Kiritani *et al.*, 1972; Nyffeler and Benz, 1988). The presence of teeth on the chelicerae (i.e. bumps and ridges on the chelicerae at the base of the fang) of spiders can be an indication of whether or not a spider species can masticate prey (Foelix, 1996). Masticating prey items is commonly used by wandering spiders including wolf and jumping spiders (Lycosidae, Salticidae) although it is also sometimes used by web-building spiders, especially if they are very hungry (Foelix, 1996). Masticating prey is thought to allow spiders to extract a greater amount of nutrients from prey items, although this remains to be tested.

Extraoral digestion is an important aspect of spider feeding that needs to be taken into account when examining nutrient intake by spiders because not all nutrients present in prey can be ingested or utilized by spiders (Cohen, 1995; Furrer and Ward, 1995; Lang and Klarenberg, 1997; Wilder *et al.*, 2010). In particular, the ratio of nutrients extracted from the body of an arthropod prey can differ significantly from the total ratio of nutrients present in a live prey. For some nutrient classes, such as lipids, this is less of a problem because it is likely that nearly all of these nutrients can be digested by the spider and ingested. However, the nitrogen or protein content of whole prey could differ significantly from what spiders are able to consume depending upon the amount of proteins that are incorporated into the exoskeleton of the insect prey during sclerotization (Andersen, 1979; Andersen *et al.*, 1995). For example, several studies have found that the concentration of nitrogen in prey remains discarded by spiders following feeding is significantly higher than the concentration of nitrogen in whole intact prey bodies (Lang and Klarenberg, 1997; Wilder and Rypstra, 2010; Wilder *et al.*, 2010). Fortunately, there is a relatively easy solution to this potential problem. It is possible to estimate the nutrients ingested by spiders when they feed on prey item by measuring the nutrient content of a whole prey item and subtracting the nutrient content of the remains that a spider discards following feeding (see Wilder *et al.*, 2010). Extraoral digestion actually facilitates the study of spider nutrition by allowing easier calculations of the efficiency with which nutrients are utilized by spiders. The quantity of nutrients that a spider ingests from a prey is equivalent to the nutrients that are assimilated and nearly all waste products are metabolic by-products (Edgar, 1971).

2.3 NUTRIENT STORAGE

Spiders may store significant amounts of lipid in their bodies. Lipid storage is strongly related to the diet of the spider with greater lipid storage when fed diets with higher lipid content (Jensen *et al.*, 2010, 2011a,b; Toft *et al.*, 2010).

For example, the lipid content in *Pardosa amentata* fairly closely follows that of their prey and both range from about 10% to 30% by dry mass (Jensen *et al.*, 2010). Significant storage of lipid is likely related to the pulsed feeding of spiders and the need to maintain sufficient stores of energy to fuel metabolic needs between prey capture events (Jensen *et al.*, 2010; Venner and Casas, 2005). However, not all spiders have high-lipid content in their bodies. For example, male spiders often have relatively small bodies and may carry low stores of nutrients to avoid being burdened by excess weight (Moya-Laraño *et al.*, 2007). The bodies of male *Hogna helluo* (Lycosidae) are only composed of 7% lipid, which is far less than many insects and juvenile spiders which have around 20% lipid in their bodies on average (Toft *et al.*, 2010; Wilder and Rypstra, 2010). The lipid content of spider bodies is important for spider nutrition because it may affect their nutritional requirements (e.g. spiders with low lipid in their body may need to consume high-lipid prey) and affects the nutritional benefit of intraguild predation and cannibalism for spiders (e.g. low-lipid male spiders compared to higher lipid juvenile spiders or insects; Wilder and Rypstra, 2010).

Spiders also have relatively high nitrogen content in their bodies (11.7%) compared to insects (herbivores: 9.5%; carnivores: 10.8%; Denno and Fagan, 2003; Fagan and Denno, 2004; Fagan *et al.*, 2002; Matsumura *et al.*, 2004), although the reason for this high nitrogen content is unclear. Fagan *et al.* (2002) suggested that high nitrogen content in carnivores, in general, compared to herbivores may result from carnivores having a bias caused by gut contents (carnivores would have higher protein food in their guts than herbivores), morphological differences related to prey capture (e.g. greater musculature which could be required to subdue prey), or sequestration or storage of nitrogen in carnivores. Gut contents alone are unlikely to explain the higher nitrogen content of spiders relative to carnivorous insects, given that both likely feed on prey with similar nutrient content. The higher nitrogen content of spiders could be related to their use of silk, which would require that significant amounts of amino acids be stored in the spider's body to be used when silk is produced (Nentwig, 1987). High nitrogen content of spiders could also be related to their use of guanine for coloration (Oxford and Gillespie, 1998). Significant amounts of guanine can be sequestered in guanocytes in the intestinal diverticula of the spider (Nentwig, 1987). These guanocyte cells play an important role in accumulating nitrogen into guanine deposits, which eventually may be excreted (Nentwig, 1987). However, some spiders may retain some guanine to use as a white pigment or possibly as a storage reserve for nitrogen (Nentwig, 1987; Oxford and Gillespie, 1998). Unfortunately, very little is known about the ability of spiders to store nitrogen and if guanine or other compounds (e.g. hexamers or silk proteins) act as storage reserves (Nentwig, 1987; Tefler and Kunkel, 1991). However, significant variation in the nitrogen content of spiders with similar total biomass (e.g. Toft *et al.*, 2010) suggests that spiders may have some method of storing nitrogen.

The ability of spider bodies to vary both in lipid and in nitrogen content has important implications for the measurement of body condition. Measures of body condition (i.e. the relative size of variable parts of the body, such as total mass or abdomen width, controlling for the fixed size of an individual, such as tibia-patella length or carapace width) are often used to assess the recent feeding status or nutritional state of spiders (Jakob *et al.*, 1996). There has been a lot of debate in the literature about which metric or type of analysis is most appropriate for measuring body condition (Green, 2001; Jakob *et al.*, 1996; Kotiaho, 1999; Moya-Laraño *et al.*, 2008; Schulte-Hostedde *et al.*, 2005; Stevenson and Woods, 2006). However, there has been relatively little discussion about what measurements of body condition actually indicate about the nutritional status of spiders and if these measures are the most appropriate way of measuring spider nutritional status. A major assumption of measurements of body condition is that lipid content of the spider is the only aspect of the spider body that varies with feeding. However, spiders can vary in both lipid and protein content and these differences in nutrient content, even when spiders are similar in overall size, could indicate dietary deficiencies or excesses and could affect spider performance and behaviour in different ways (Jensen *et al.*, 2010, 2011b; Toft *et al.*, 2010). Hence, in some cases, measures of body condition may not provide enough information about spider nutritional status and should be supplemented by randomly sacrificing individuals from each treatment group for more in-depth analyses of body nutrient content.

Finally, the consequences of variation in spider body composition for spider nutrition are not always clear. Fagan and colleagues (Denno and Fagan, 2003; Fagan and Denno, 2004; Fagan *et al.*, 2002; Matsumura *et al.*, 2004) argued that the high nitrogen content of spiders and other carnivorous arthropods relative to their herbivorous prey indicated that carnivores were nitrogen limited in nature based on the assumption that individuals need to eat food with nutritional content similar to that of their own body. However, alternate views would suggest that the high nitrogen content of spider bodies indicates lipid limitation either because it indicates spiders have excess nitrogen to store in their bodies or because the high nitrogen content is a result of low-lipid content (i.e. the percentage of lipid and nitrogen in an individual's body are interdependent; Raubenheimer *et al.*, 2009; Wilder *et al.*, 2010). Further experiments are needed to examine if spiders bias their feeding towards particular nutrients when taken from the field or have particular shifts in their body composition after feeding in the laboratory similar to assays used by Raubenheimer *et al.* (2007) for ground beetles.

2.4 SILK

One of the defining features of spiders is the production of silk (Foelix, 1996; Herberstein *et al.*, 2000; Nentwig, 1987; Vollrath and Selden, 2007). Although a number of other insects produce silk during various life stages for specific purposes (e.g. cocoons), spiders have mastered the use of silk for a wide range of functions

(Foelix, 1996; Nentwig, 1987; Vollrath and Selden, 2007). Much is known about the production of silk by spiders, given its interesting physical properties and its potential economic applications (Fu *et al.*, 2009; Porter and Vollrath, 2009). In this section, I will discuss the basic composition of silk and how variation in the production of silk can affect the nutritional requirements of spiders. More detailed reviews of the biochemistry and physical properties of silk and the anatomy and physiology of the various types of silk glands can be found in Nentwig (1987), Foelix (1996), Vollrath and Selden (2007) and Fu *et al.* (2009).

Silk is used by different taxa of spiders in a variety of ways. Ecologically, spiders can be divided into two major groups: wandering and web building (Foelix, 1996). Wandering spiders (e.g. Lycosidae, Salticidae, Oxyopidae, and most other members of the RTA clade) primarily use silk for basic functions including building retreats, encasing eggs and as draglines that are periodically anchored to the substrate to protect against falling (Foelix, 1996; Nentwig, 1987; Vollrath and Selden, 2007). Web-building spiders use silk for these same basic functions but also build some type of prey capture structure with silk. Spider webs can vary tremendously in size and structure and include, but are not limited to, simple nets that are wrapped around prey, tangle-webs of haphazardly strung silk, horizontal sheet webs, two-dimensional orb webs and three-dimensional ''dome'' webs (Blackledge *et al.*, 2009; Vollrath and Selden, 2007). In addition, while some of these webs are fairly basic in structure, others can contain elaborate ornamentation, which may be important for strengthening the web, preventing damage from larger animals and increasing foraging success (Herberstein *et al.*, 2000). Prey capture webs vary tremendously in their investment of silk from a small amount (e.g. webs of net casting spiders; Deinopidae) or a moderate amount that can be eaten and recycled (e.g. some orb webs; Araneidae) to a large amount that cannot be recycled (e.g. sheet webs; Agelenidae) (Blackledge *et al.*, 2009; Vollrath and Selden, 2007). For example, *Araneus diadematus* (Araneidae) produce webs that contain 0.1 mg of silk that they consume and recycle when building new webs, while *Agelena limbata* can produce webs 100 times that weight (10 mg) that are not recycled (Peakall, 1971; Tanaka, 1989). The quantity of silk used by a given spider may be an important factor affecting its nutritional requirements, especially requirements for protein.

Protein is the major constituent of spider silk (Foelix, 1996; Nentwig, 1987). The amino acid composition of this protein is fairly conserved among species and even among silk produced by different silk glands (Craig *et al.*, 1999, 2000; Foelix, 1996; Nentwig, 1987). The vast majority of spider silk (e.g. about three-quarters of the amino acid content) is made up of five non-essential amino acids: alanine, glycine, proline, glutamic acid and serine (Andersen, 1970; Craig *et al.*, 1999, 2000; Foelix, 1996; Lombardi and Kaplan, 1990; Tillinghast and Christenson, 1984; Xu and Lewis, 1990). In comparison, 86% of the silk of the silkworm, *Bombyx mori*, is made up of only three amino acids: alanine, glycine and serine (Craig *et al.*, 1999; Lucas *et al.*, 1958; Nentwig, 1987). Essential amino acids only comprise approximately 5% or less of the total amino acid content of spider silk (Andersen, 1970; Foelix, 1996; Lombardi and Kaplan, 1990). Hence, dietary essential amino acids are unlikely to limit silk production in spiders and most amino acids in silk can presumably be synthesized from other amino acids if they are not present in high enough concentrations in the diet.

Spiders have a constant need to produce silk. Wandering spiders continually deposit silk as they move and web-building spiders need to repair and replace webs that become damaged (Breed *et al.*, 1964; Foelix, 1996). The silk-producing glands of spiders continue to be replenished as they are depleted and there seems to be little feedback to prevent the glands from sequestering amino acids from the body to produce silk (Breed *et al.*, 1964; Peakall, 1966; Witt, 1963). Even if spiders are starved, they will continue to replenish the contents of their silk glands as silk is removed to the point of losing half of their body mass (Witt, 1963). Given the high-protein cost of silk production, spiders have evolved several mechanisms to mitigate this cost. One mechanism is plasticity in silk use or web production. At moderate hunger levels, spiders may increase investment in webs to build larger webs that have a greater probability of capturing prey (Sherman, 1994). However, under extreme levels of food deprivation or silk depletion, spiders may reduce the size of their web or reduce investment in other silk containing structures such as silk-lined burrows (Eberhard, 1988; Walker *et al.*, 1999). Spiders also make allocation decisions between web building and other activities that require nutrients. For example, in the orb-weaving spider *Larinioides cornutus* (Araneidae), well-fed females reduce the size of their web and, instead, shift their protein allocation to building eggs while food-limited females may invest less in eggs and more in larger webs (Sherman, 1994).

Web recycling is an important mechanism that can be used to reduce the cost of silk use by web-building spiders (Opell, 1998). Web recycling is when spiders ingest their current web to gain protein needed to produce a new web (Opell, 1998; Peakall, 1971). This behaviour truly is recycling and is very efficient (Opell, 1998; Peakall, 1971). For example, amino acids that are ingested from silk can be found in the silk glands of a spider within 30 min, indicating that these amino acids are efficiently routed for subsequent use as silk (Peakall, 1971). In addition, radioactive labelling experiments have shown that this recycling is 92–96% efficient with the remaining small percentage of amino acids allocated to the body of the spider and almost none of these amino acids are broken down and excreted (Peakall, 1971). Web recycling is also beneficial because it allows spiders to ingest nutrients from very small sources of food that are stuck to the web (e.g. pollen or tiny insects) and that would not normally be consumed (e.g. Smith and Mommsen, 1984). In addition to spiders recycling their own webs, some kleptoparasitic spiders will sometimes feed on the silk in their host spider's web (Nentwig, 1987; Tso and Severinghaus, 1998). While silk recycling seems like a very efficient nutritional strategy, not all web-building spiders recycle their webs. Web recycling seems most common in the family Araneidae but is rare in other closely related families within the Orbiculariae superfamily and absent in a range of more distantly related families outside of this clade (Coddington, 2005; Opell,

1998). Further research is needed on the phylogenetic distribution of web recycling and why it is present in some species and absent in others.

In addition to silk, which provides a structural framework for webs, some spiders also incorporate compounds in ''sticky droplets'' onto parts of their webs (Blackledge *et al.*, 2009; Foelix, 1996; Nentwig, 1987). More ''primitive'' spider webs consist exclusively of silk and often contain very fine mesh of silk produced by a cribellar plate that is used to entangle prey (Foelix, 1996; Opell, 1997, 1998). Over evolutionary time, these cribellate webs were replaced by webs with sticky droplets of adhesive chemicals that help hold a prey item in a web (Blackledge *et al.*, 2009; Opell, 1997, 1998). Webs with sticky droplets require significantly less silk than cribellate webs and may be more effective at holding prey (Heiling and Herberstein, 2000; Opell, 1998). The inclusion of sticky droplets in the spider web was a major evolutionary advance that was accompanied by significant diversification of spiders (Blackledge *et al.*, 2009; Opell, 1997, 1998).

Approximately half of the weight of webs of some orb-weaving spiders is non-silk components associated with the sticky droplets (Nentwig, 1987; Opell, 1997). Of these non-silk components, approximately 80% is water, which is critical for maintaining the adhesive properties of the sticky droplets and the elasticity of the silk in the sticky spiral part of the web (Gosline *et al.*, 1986; Nentwig, 1987; Opell, 1997). The rest of the non-silk components of the spider web consist of a range of chemicals including, but not limited to, inorganic salts (e.g. KNO_3, KH_2PO_4, which assist in attracting and retaining water), glycoproteins (which act as the adhesive glue) and a range of low-molecular weight compounds (e.g. glycine, alanine, *N*-acetyltaurine, putrescine, GABamide, glycine betaine and choline) (Anderson and Tillinghast, 1980; Foelix, 1996; Nentwig, 1987; Tillinghast and Christenson, 1984; Towley *et al.*, 1991; Townley *et al.*, 2006; Vollrath *et al.*, 1990). The concentrations of these chemicals in the sticky droplets are not constant and can vary both within and among species (Anderson and Tillinghast, 1980; Higgins *et al.*, 2001). Some of these chemicals appear to be dietary essential nutrients, especially choline, and their inclusion in the sticky droplets can be affected by general food limitation or the specific availability of these chemicals in a spider's diet (Higgins and Rankin, 1999; Townley *et al.*, 2006). Little is known about the functional consequences of changes in the biochemical composition of sticky droplets, but it likely affects their overall effectiveness, the rate at which their effectiveness degrades over time or the overall ability of the sticky droplets to withstand environmental conditions (e.g. low humidity) (Higgins and Rankin, 1999; Higgins *et al.*, 2001; Nentwig, 1987; Townley *et al.*, 2006).

2.5 METABOLIC RATE

Spiders have very low resting metabolic rates compared to other animals. On average the resting metabolic rate of spiders is 70% of that of comparable ectothermic arthropods (Anderson, 1970; Greenstone and Bennett, 1980;

Prestwich, 1983a,b). The low resting metabolic rate of spiders may be due to the fact that they use hydrostatic pressure for extending their appendages (i.e. leg musculature is primarily used for flexion) and may be able to maintain a constant posture by maintaining constant hydrostatic pressure with a few small muscles instead of constantly using their leg musculature (Anderson and Prestwich, 1975; Carrel and Heathcote, 1976; Parry and Brown, 1959). The low metabolic rate of spiders is ecologically important because it may allow them to tolerate variation or unpredictability in their food supply and is nutritionally important because it suggests that spiders may be very efficient in their use of energy.

The low resting metabolic rate of spiders may be an important factor that allows them to survive extended periods of time without food (Anderson, 1974; Nentwig, 1987; Tanaka and Ito, 1982). In addition, spiders can further reduce their metabolic rate below their already low levels when they experience periods of food limitation (Anderson, 1974; Ito, 1964; Miyashita, 1969; Stoltz *et al.*, 2010; Tanaka and Ito, 1982). Reductions in metabolic rate of 16% have been observed after only a couple of days of starvation (Ito, 1964). For most spiders that have been examined, metabolic rate steadily declines over the first two weeks of starvation after which it remains relatively constant (Anderson, 1974; Ito, 1964; Miyashita, 1969; Stoltz *et al.*, 2010; Tanaka and Ito, 1982). Resting metabolic rates during periods of starvation are typically less than half the basal metabolic rate of well-fed spiders (Anderson, 1974; Ito, 1964; Miyashita, 1969; Philip and Shillington, 2010; Stoltz *et al.*, 2010; Tanaka and Ito, 1982). A low basal metabolic rate and an incredible ability to further reduce metabolic rate may explain the ability of some spiders to survive exceptionally long periods without food, which are sometimes in excess of 200 or 300 days (Anderson, 1974; Nentwig, 1987; Tanaka and Ito, 1982).

Spiders may elevate their metabolic rate during a variety of activities including searching for food, producing webs, producing eggs or courtship. However, even when spiders elevate their metabolic rate during activities, the increase in respiration and metabolic rate is relatively low compared to other animals (e.g. two to six times resting rates; Anderson and Prestwich, 1982; Ford, 1977a,b; Prestwich, 1983a,b). One explanation for low respiration rates during activity is that spiders rely heavily on anaerobic metabolism to fuel their muscles (Linzen and Gallowitz, 1975; Prestwich, 1983a,b; Prestwich and Ing, 1982). Spider leg muscles are more similar to the hind jumping legs of *Locusta* grasshoppers than other insect legs in that there are very few or no mitochondria in the muscle fibres of spiders (Linzen and Gallowitz, 1975). In addition, spider muscles are also comparable to those of grasshoppers because they perform best under short rapid bouts of activity and become exhausted relatively quickly (Linzen and Gallowitz, 1975; Prestwich, 1988; Wilson and Bullock, 1973). Rapid exhaustion of spiders is often not a problem because many spiders, including both web-building and ''wandering'' spiders, are sit-and-wait predators that spend very little time in active locomotion (Ford, 1977a,b; Prestwich, 1977). For example, in the wolf spider *P. amentata*, the daily energy loss attributed to locomotion was estimated to be only 1% of the daily energy usage of spiders (Ford, 1977a). Web-building spiders may increase their metabolic rate almost 10–20 times above basal (i.e. resting) levels while building a web (Ford, 1977a,b; Peakall and Witt, 1976; Prestwich, 1977, Tanaka, 1989). However, the energetic costs of web production are likely to be relatively small, especially compared to the protein costs of silk, because web building is often a relatively short process and some spiders are able to recycle web proteins, which can substantially reduce the metabolic cost of silk production (Ford, 1977a,b; Peakall and Witt, 1976; Prestwich, 1977; Tanaka, 1989).

Activities associated with reproduction may require significant increases in metabolic rate. These costs may also be partitioned differently among the sexes. For example, in the wolf spider *Pardosa lugubris*, females invest 73% of their ingested energy in respiration and 26% in reproduction, while males invest 81% of their ingested energy in respiration and only 16% in production (Edgar, 1971). The higher investment of energy by females into production is likely due to their investment of significant amounts of lipid into their eggs, while males only need to invest in the production of sperm (i.e. besides the common production costs of repairing and maintaining their bodies and producing silk draglines) (Edgar, 1971; Kessler, 1971; Wilder and Rypstra, 2010). The differences in respiration rate between the sexes is likely related to the high energetic needs of males for agonistic encounters with competing males and for courting females (Kotiaho *et al.*, 1998; Lomborg and Toft, 2009; Riechert, 1988; Uetz *et al.*, 2002; Watson and Lighton, 1994). Hence, the reproductive state (i.e. juvenile vs. adult) and sex of spiders may be important factors to take into account when examining nutritional requirements and performance on different diets.

3 The food of spiders in nature

One of the first steps to understanding the nutritional ecology of spiders is to examine the types and amounts of nutrients that are ingested by spiders in nature. To gain a more complete understanding of spider feeding, it is important to integrate studies on multiple scales and disciplines ranging from an ecological perspective that considers variation in prey taxa captured by spiders to a more physiological level that considers the biochemical composition of those prey and what nutrients are actually ingested by spiders. In this section, I consider three aspects of spider feeding in nature. First, I discuss evidence that prey availability for spiders is highly variable and often limited and the implications this has for studying spider nutrition. Second, I review studies that have documented the types of prey that are captured by spiders in nature and the consequences of prey capture for nutrient intake by spiders. Finally, I discuss prey and nutrient choice and how this may allow spiders to achieve an optimal intake of nutrients.

3.1 FOOD LIMITATION AND VARIABILITY

Spiders, and carnivores in general, have long been hypothesized to be food limited in nature (Hairston *et al.*, 1960; White, 1978; Wise, 1993, 2006). Evidence for food limitation in spiders is reviewed in Riechert and Harp (1987), Nentwig (1987) and Wise (1993, 2006). Hence, I only provide a brief overview of this evidence before discussing the implications of overall food limitation for spider nutrition.

A wide range of evidence supports the idea that the overall quantity of prey available to spiders is limited in nature. Observational studies of spiders in the field have shown that many species have very low capture rates of prey (Edgar, 1969; Nyffeler and Breene, 1990). These rates are much lower than the rates at which spiders can consume prey in the laboratory and the body condition of spiders in the field is also lower than well-fed spiders in the laboratory (Anderson, 1974; Hardman and Turnbull, 1974; Kessler, 1971; Miyashita, 1968; Wilder and Rypstra, 2008a). Further, supplementing the prey available to spiders in the field often results in large increases in growth, reproduction and/or survival in a number of families: Araneidae (Spiller, 1984; Wise, 1979, 1983), Eresidae (Lubin and Henschel, 1996), Linyphiidae (Wise, 1975), Lycosidae (Moya-Laraño *et al.*, 2003) and Pisauridae (Kreiter and Wise, 2001). Prey supplementation also increases the aggregation and population density of spiders in the field (Olive, 1982; Rubenstein, 1987; Rypstra, 1985; Wise, 1993). Using a novel technique, Bilde and Toft (1998) also were able to estimate the degree of starvation of spiders in the field by comparing the consumption rate of recently captured spiders with reference spiders that had been starved for known periods of time. In two species of spiders, the hunger level of individuals in the field was equivalent to at least 1 week of starvation (Bilde and Toft, 1998). Hence, abundant evidence suggests that overall food limitation is a frequent occurrence for spiders.

While spiders may frequently experience food limitation in nature, there may also be periods of time where plenty of prey are available. At a small scale, some individual prey items may be too large to be consumed by spiders in a single feeding bout (Rypstra and Tirey, 1991), even though spiders have developed feeding mechanisms to maximize nutrient ingestion and minimize the ingestion of indigestible bulk (i.e. extraoral digestion; Cohen, 1995). There can also be incredible temporal or spatial variation in the quantity of prey available to spiders due to variation in the phenology, activity or habitat preferences of insect populations (Cherrett, 1964; Henschel *et al.*, 2001; Kato *et al.*, 2003; Miyashita, 1992; Nentwig, 1983; Rypstra, 1985; Venner and Casas, 2005). For example, pulsed emergences of aquatic insects on a scale of a couple weeks or months can increase the abundance of insect prey available to spiders 2.6- to 22.9-fold in riparian areas compared to areas 30–60 m away (Henschel *et al.*, 2001). While pulses of food or large prey items may be quite infrequent, they can be extremely important for spider growth, survival and reproduction (Venner and Casas, 2005).

In the study of spider nutrition, a major general research goal is to understand how the concentrations of nutrients in prey affect the performance of spiders and if there is an optimal concentration of nutrients at which spider performance (e.g. growth, reproduction, survival) is maximized. The overall quantity of food available to spiders is important for guiding future research to examine the implications of these nutrients for spiders in nature. When food is overall limited in availability, the nutritional content of each prey item can have a large effect on the performance of spiders and individuals may be constrained to particular diets with little opportunity for selectivity (Mayntz *et al.*, 2003). This suggests that, when prey is limited, nutritional research should focus on how spiders perform when constrained to particular diets compared to diets with optimal concentrations of nutrients. In contrast, when food is overabundant, spiders may have the opportunity to be selective of their diet (Riechert, 1991). For these situations, it may be important to study if and how spiders select an optimal concentration of nutrients. Spiders could be selective of their diet at a range of scales from different parts of an individual animal to major prey taxa (Mayntz *et al.*, 2005).

3.2 COMPOSITION

The use of a wide range of techniques (e.g. direct observation, genetic and ELISA analyses of gut contents) has allowed investigators to collect an incredible amount of data on the taxa of prey captured by spiders in the field (Agusti *et al.*, 2003; Greenstone, 1977; Greenstone and Shufran, 2003; Harwood *et al.*, 2004, 2007). Yet despite decades of research on prey capture and feeding by spiders (Nentwig, 1986, 1987; Nyffeler, 1999; Robinson and Robinson, 1970), there is very little information available on nutrient ingestion by spiders in the field. In this section, I summarize what is known about prey capture by spiders in the field, describe why this has done little to advance our understanding of spider nutrition and highlight some of the challenges involved with quantifying nutrient intake by spiders in the field.

Some general patterns have emerged from studies of spider diets. Overall, the most common taxa of prey consumed by spiders are Diptera and Hemiptera (Nentwig, 1983, 1985, 1986, 1987; Nyffeler, 1999). Hymenoptera can also account for a significant proportion of the diet of some spiders (e.g. 10–14% on average; Nyffeler, 1999). However, the type of Hymenoptera consumed and their frequency in spider diets varies widely among spider species (Denne, 1982; Nentwig, 1986, 1987; Nyffeler, 1999). Lepidoptera are rarely consumed by spiders most likely because the scales on their wings prevent them from sticking to spider webs and because some are unpalatable and are released by spiders (Nentwig, 1982, 1983; Vasconcellos-Neto and Lewinsohn, 1984). Coleoptera are also infrequently consumed by spiders presumably because their high amount of sclerotization and protective elytra make them more difficult to subdue (Nentwig, 1983).

Despite general patterns, there is significant variation in the taxa of prey consumed by spiders, likely due to variation in the relative abundances of prey taxa (e.g. spatial and temporal), the stochasticity of prey capture and variation in the habitat use and hunting strategies of spiders (Fritz and Morse, 1985; Henschel *et al.*, 2001; Morse, 1981; Nentwig, 1987; Nyffeler, 1999). In addition to variation in the frequency of different prey taxa, there can be substantial variation in the relative contribution of prey to a spider's diet, given the large variation in the size of prey items (Venner and Casas, 2005). For example, in *Argiope argentata*, small hymenopterans (0.015 g each) represented 62% of the prey items captured but only 17% of the prey biomass, while larger prey such as orthopterans represented only 11% of the prey items captured but over 55% of the prey biomass (Robinson and Robinson, 1970). Given widespread variation in prey capture by spiders, it may be difficult to make generalizations about spider diets.

Regardless of consistency or variation in the taxa of prey captured, one major question that remains unclear is whether or not prey taxonomy is a reliable predictor of the nutrient content present. There may be general differences in the nutrient content of some prey taxa. For example, phylogeny has some ability to predict the lipid content of insects (Lease and Wolf, 2011). Certain prey taxa may also be more likely to have toxins than other prey taxa (Silva and Trigo, 2002). In addition, a number of studies have also found that carnivorous insects have higher nitrogen content than herbivorous insects, although the biochemical allocation of this nitrogen (e.g. sclerotization of the cuticle vs. protein) is not known (Denno and Fagan, 2003; Fagan and Denno, 2004; Fagan *et al.*, 2002; Matsumura *et al.*, 2004; Wilder and Eubanks, 2010). Yet other evidence suggests that there can be substantial variation in the nutritional content within a single species of prey. For example, changing the nutrient content of the diet on which a single species of prey is reared can have a large effect on the growth, reproduction and survival of spiders consuming those prey (Table 2; discussed in more detail in Section 4). In addition, a single species of prey can vary in the amount of lipid and protein available for spiders to ingest depending upon the feeding level, sex and developmental stage of the prey (Wilder *et al.*, 2010). This intraspecific variation in nutrient content can actually be quite significant and be almost as large as interspecific variation in lipid and protein content (Wilder *et al.*, 2010). Further research is needed to examine the scales (e.g. individuals within a species, species, families) at which prey vary in nutritional content as this is critical for connecting prey capture with the concentrations of macronutrients that spiders ingest in nature. Small scale variation in the nutrient content of prey (i.e. variation among individuals of the same species) does not preclude the study of spider nutrition in the field, but it would make it much more difficult to study.

3.3 PREY CHOICE

A major theme in the study of nutritional ecology is that animals have specific macronutrient requirements for optimal growth and that, when given a choice, they often select food with the concentration of nutrients that meets these requirements (Lee *et al.*, 2008; Maklakov *et al.*, 2008; Simpson and Raubenheimer, 1993, 1995; Simpson *et al.*, 2004). However, the degree to which spiders are selective of prey remains unclear. Some authors have argued that spiders are indiscriminate while others have argued that they can be quite selective (Toft, 1999; Wise, 2006). The wide variation in the nutritional content of prey and the large effect of prey nutrients on spider performance (Table 2) suggest that spiders could benefit from selecting prey based on their nutrients. There are three major questions related to nutrition and prey choice in spiders: (1) Do spiders choose prey based on their nutrient content? (2) How (or at what scale) do spiders choose prey or nutrients? (3) What are the consequences of prey choice for arthropod communities?

Some evidence suggests that spiders may choose prey based on their nutrient content. In the wolf spider, *H. helluo*, females that were starved for 1 week were hesitant to engage in sexual cannibalism but would readily consume a cricket the same size as a male (Wilder and Rypstra, 2010). Hesitancy to cannibalize males could be due to the very low lipid content of male bodies (5% lipid) and the much closer match of crickets (20% lipid) and an average insect (24% lipid) to the concentrations of lipids that female spiders need to produce eggs (25% lipid; Wilder and Rypstra, 2010). In a more controlled study of prey choice, Jackson *et al.* (2005) found that the jumping spider, *Evarcha culicivora*, preferentially preyed upon female mosquitoes that had recently fed upon blood over similarly sized mosquitoes that had recently fed on sugar water. By manipulating the content (i.e. blood vs. sugar) of a single species of prey, Jackson *et al.* (2005) were able to control for potentially confounding effects of prey species (e.g. behaviour, defences) and show that the spiders were choosing prey based on the nutrients in the prey body. Future studies are needed to test if spiders are able to discriminate between prey that vary in a variety of nutrients, including general nutrient manipulations (e.g. high vs. low nutrient) or even specific macronutrient classes (e.g. lipid vs. protein; Jensen *et al.*, 2011a,b).

In addition to selecting particular prey items, spiders may exercise selectivity at the scale of a single prey item. Particular parts of an insect body may contain higher concentrations of lipid (e.g. the fat body in the abdomen) or higher concentrations of protein (e.g. muscles in the leg or thorax). In the specialist ant-eating jumping spider, *Zodarion rubidium*, individuals preferentially foraged on the foreparts of ants (Pekár *et al.*, 2010). When fed the foreparts and gaster separately or together, spiders grew faster when they were fed two ant foreparts than when they were fed either two gasters or a forepart and a gaster, suggesting that the spiders adaptively chose the parts of prey that had the most appropriate concentrations of nutrients to support growth (Pékar *et al.*, 2010). A web-building spider, *Stegodyphus lineatus*, was also able to selectively extract nutrients from individual prey items and extracted a greater amount of lipid from a prey item when the spider had previously been maintained on high-protein prey and vice versa (Mayntz *et al.*, 2005). Wolf spiders also appear to regulate their intake of nutrients from individual prey. However, since wolf spiders feed by masticating prey, they seem to regulate nutrient ingestion by consuming different overall quantities of prey that vary in average nutrient content instead of specifically targeting parts of prey that vary in nutrients (Jensen *et al.*, 2011b; Mayntz *et al.*, 2005). Feeding mode (i.e. sucking prey nutrients out through a small hole in the prey exoskeleton or masticating prey) may be an important consideration when examining the ability of spiders to selectively extract nutrients from prey. The ability of spiders to select specific nutrients within an individual prey item is important to recognize because it indicates that spiders may still be regulating their diet even if they do not selectively kill particular prey species.

Spider prey choice has important implications for arthropod community dynamics in both natural and agricultural landscapes. In particular, the study of spider nutrition may provide an ideal framework to integrate among a wide range of scales in biology from individual physiology to community or ecosystems dynamics. The nutritional requirements of spiders (which may vary among species or life stages) may influence the species of prey that are killed and consumed and, in turn, affect the composition of arthropods in communities. There could also be dynamic feedback between spider physiology and community composition. For example, environments with a high availability of high-lipid prey could result in protein-limited spiders that subsequently engage in higher rates of cannibalism or intraguild predation on other spiders, which have low-lipid and high-protein content. Given the recent interest in connecting the behaviour and physiology of individuals with community dynamics (Beckerman *et al.*, 2010; McGill *et al.*, 2006; Simpson *et al.*, 2009), spider nutritional ecology may prove to be a very promising area of research.

3.4 CONCLUSIONS: WHAT DO SPIDERS EAT?

We know a lot about the frequency and types of prey that spiders capture and feed on but remarkably little about the concentrations of nutrients that spiders ingest. The identity of the prey consumed by spiders is critical because spiders consume live prey that can vary in a range of characteristics including nutrients, morphological and chemical defences (e.g. hardening of the exoskeleton), weapons and behaviour. However, regardless of the packaging, all prey are ultimately sacs of nutrients and it is the nutrients that are removed from prey, assimilated into the spider body and used for maintenance, growth and reproduction. Identifying what nutrients spiders are ingesting in nature, how nutrient ingestion varies (e.g. by habitat, season, prey capture strategy) and if or how spiders regulate their feeding to maintain a particular intake of nutrients is the first major goal for gaining a thorough understanding of the nutritional ecology of spiders.

4 Nutrition and spider performance

Once we have data on the intake of nutrients by spiders in nature, the next obvious question to address is how this variation in nutrient intake affects spider performance (e.g. behaviour, development time, development rate, size at maturity, reproduction and survival). In this section, I review studies of how spider nutrition affects performance with studies categorized by the scale at which they have studied spider nutrition, including manipulations of (1) different prey species or combinations of prey, (2) groups of nutrients (e.g. simultaneous addition of nutrients from several nutrient classes) and (3) specific nutrient classes (Table 2). These categories represent a continuum of understanding spider nutrition from a practical and ecological perspective (different prey species) to a functional and physiological perspective (individual nutrient classes). While these categories are separated for logistical reasons, a thorough understanding of spider nutrition requires integration among levels from nutrient classes to prey species and integration of these performance studies with prey capture and nutrient intake of spiders in the field.

4.1 PREY SPECIES

A number of studies have documented that the species of arthropod prey fed to spiders can have a strong effect on the performance of spiders (Table 2). Spiders fed particular species of prey have higher survival, growth and reproduction than spiders fed other species of prey (Table 2). This is important from an ecological perspective because it indicates that variation in the types of prey consumed by spiders in nature (e.g. Section 2.2: Composition) can have important consequences for spider fitness. However, as will be discussed in this section, studies of how different prey species affect spider performance are only a starting point for understanding spider nutrition and need to be combined with more controlled manipulations to reveal which aspects of prey (e.g. prey morphology, toxins or nutrients) are responsible for the observed effects on spider performance.

Studies of spider performance when fed different species of prey differ widely in the insights that they provide to spider nutrition. Differences in the performance of spiders fed two different species of prey could result from differences in the concentration of one or more nutrients, toxins, behaviour or morphological defences (Matsumura *et al.*, 2004). Even prey that are taxonomically similar can have very different effects on the performance of spiders. For example, spider performance varies dramatically depending upon the species of Collembola on which they feed (Marcussen *et al.*, 1999; Toft and Wise, 1999). Wolf spiders (*Pardosa* spp.) often die fairly quickly and grow very little when feeding on Collembola from the genus *Folsomia* but have very high survival and rapid growth when fed other Collembola including those from the genus *Tomocerus* (Rickers *et al.*, 2006; Toft and Wise, 1999). Spiders also have high levels of survival and growth when fed *Isotoma anglicana* but will die faster than spiders that are starved when fed congeneric *Isotoma trispinata* (Marcussen *et al.*, 1999; Toft and Wise, 1999). Measuring a few aspects of the chemical composition of prey does not necessarily resolve the problem because it would be difficult to measure all potential differences in prey biochemistry and, even if this were done, the results would only provide a correlative link between prey nutrients and spider performance.

Studies that examine the response of spiders to single or mixed prey species diets can provide more insight into the importance of nutrition for spider performance. The performance of spiders fed a combination of prey (mixed diets) can be higher than, equal to, or lower than the performance of spiders fed the single species of prey that resulted in the highest performance of spiders on a monotypic diet. A synergistic effect of diet mixing (higher performance when fed a mix than the single best prey) suggests that the prey included in the combination are insufficient on their own but complementary (Bilde and Toft, 2000; Sigsgaard *et al.*, 2001; Toft, 1995, 1999; Toft and Wise, 1999; Uetz *et al.*, 1992). Prey could be complementary if each contained a different dietary essential nutrient that was lacking in the other prey. Prey could also be complementary if the ratio of nutrients present in each prey was skewed relative to the needs of the spider but was similar to the spider's needs when the prey were combined (e.g. if one prey was lipid rich and the other was protein rich but the spider needed an even balance of lipid and protein). In other cases, the performance of spiders fed a mixed diet is similar to that of spiders fed the single best prey item. This pattern suggests that a single prey contains all of the essential nutrients necessary for spider survival, growth or reproduction. Finally, lower performance of spiders on mixed compared to single species diets is often attributed to the inclusion of toxic prey in the diet and the overwhelming effect of prey toxins on spider performance (Fisker and Toft, 2004; Marcussen *et al.*, 1999; Oelbermann and Scheu, 2002a; Rickers *et al.*, 2006; Toft and Wise, 1999). Lower performance on mixed diets can also indicate that a spider is specialized for a specific prey item, such as the specialist araneophagic spider, *Portia fimbriata*, where individuals fed a diet of spiders and insects had lower survival and growth than spiders fed only spiders (Li and Jackson, 1997). However, more careful measurements and manipulations are needed to test these speculative hypotheses for why spider performance differs when fed different monotypic or mixed prey diets.

The presence of toxins is one of the few aspects of prey that can be readily identified from studies of spiders fed different prey species. Evidence of toxins in prey is provided by studies where either there is little growth and high mortality when fed that prey item or, even more extreme, if consumption of a given prey results in spiders dying faster than when spiders are starved. Toxic prey can result in low growth and high mortality in mixed diets, regardless of how suitable other prey are for spider growth and survival, and some spiders appear unable to discriminate among toxic and non-toxic prey when they are provided together (Fisker and Toft, 2004; Marcussen *et al.*, 1999; Oelbermann and Scheu, 2002a; Rickers *et al.*, 2006; Toft and Nielsen, 1997; Toft and Wise, 1999). Unfortunately, very little is known about the identity of the chemicals responsible for the toxicity of particular insect prey (Nielsen and Toft, 2000). What is known is that these chemicals appear to cause higher mortality and lower growth of spiders by increasing respiration rates, decreasing absorption efficiencies or both (Fisker and Toft, 2004; Toft and Nielsen, 1997). In other words, although spiders may ingest levels of nutrients that are typically sufficient for growth, the presence of toxic compounds may prevent spiders from absorbing and assimilating the nutrients or cause spiders to metabolize the nutrients before they can be used for growth (Fisker and Toft, 2004; Toft and Nielsen, 1997). Much more research needs to be conducted on toxins in the prey of spiders because of their potential importance for spider nutrition. It is likely that the occurrence of toxic chemicals in prey is underappreciated because current growth assays are only able to detect toxic chemicals that are lethal or have a large effect on spiders but may not detect toxins that occur at lower concentrations in prey or that have smaller sublethal effects.

4.2 GENERAL NUTRIENT MANIPULATIONS

Greater evidence of the importance of nutrition for spider performance has been provided by studies that have conducted manipulations of spider diets independent of prey species. One way to manipulate the nutrient content of diets without changing the species of prey fed to spiders is to supplement prey with non-prey sources of food. A number of species of spiders will feed on nectar and other high-carbohydrate liquid foods (Jackson *et al.*, 2001; Meehan *et al.*, 2009; Pollard *et al.*, 1995; Taylor and Bradley, 2009; Vogelei and Greissl, 1989). Among wandering spiders, members of at least four families (Anyphaenidae, Miturgidae, Thomisidae and Oxyopidae) regularly feed on nectar in nature, especially adult spiders of which 25–40% may have detectable remains of nectar in their digestive tract (Taylor and Pfannenstiel, 2008). Feeding on nectar or specific carbohydrates (e.g. sucrose) results in greater survival of juveniles in several species of spiders both when carbohydrates are provided alone and as a supplement to arthropod prey (Taylor and Pfannenstiel, 2009; Vogelei and Greissl, 1989). Taylor and Pfannenstiel (2009) conducted one of the most extensive studies of non-prey food by experimentally supplementing the diet of insect prey of juvenile and adult spiders, *Cheiracanthium mildei* (Araneae; Miturgidae), with nectar. Individuals provided with nectar supplements had higher survival, faster growth and greater egg production, especially at low prey availability (Taylor and Pfannenstiel, 2009). However, this techniques has limitations as there are few non-prey foods that spiders feed on in nature (e.g. mainly pollen and nectar), these non-prey foods only span a limited range of nutrient concentrations and only a limited number of spider species have been observed to use non-prey supplements.

Another way to alter the nutrient content of spider diets is to manipulate the quantity or ratio of nutrients present in a single species of prey. Mayntz and Toft (2000, 2001) pioneered this method by raising flies, *Drosophila melanogaster*, on a low-nutrient media (i.e. potato flakes) supplemented with a wide range of nutrients. Changing the nutrient content of the media on which larval flies,

D. melanogaster, develop changes the nutrient content in the bodies of adult flies and affects the survival and growth of spiders that feed on these flies (Mayntz and Toft, 2000, 2001; Mayntz *et al.*, 2003). *D. melanogaster*, in particular, appear quite flexible in the nutrient composition of their body and a number of nutrient types can be elevated by adding them to the larval diet (Draper *et al.*, 2000; Markow *et al.*, 1999; Mayntz and Toft, 2001; Shreve *et al.*, 2007). Similar manipulations of prey nutrients by feeding prey different diets have also been conducted using house flies (*Musca domestica*; Blamires *et al.*, 2009), caterpillars (*Manduca sexta*, *Vanessa cardui*, *Junonia coenia*; Strohmeyer *et al.*, 1998) and Collembola (*Heteromurus nitidus*; Pollierer *et al.*, 2010). One way that manipulations of prey diet can affect predator nutrition is if some of the prey food is present in its gut when it is eaten by a spider (i.e. ''gut-loading''; Finke, 2003). Prey diets can also affect prey nutrient content if they change storage reserves of nutrients (e.g. fat bodies) or investment in structures in the body of the prey (e.g. muscles or eggs). Mayntz and Toft (2001) found that *D. melanogaster* reared on high-nutrient media resulted in greater survival and growth of spiders, regardless of whether the flies were recently fed or starved to clear their gut contents. Regardless of how these manipulations change the nutrients present in prey, they provide a way to examine the effects of prey nutrient content on spiders without the potentially confounding effects of prey species identity (e.g. variation in unknown compounds, prey behaviour, prey chemical and morphological defences).

Manipulations of the diet on which a prey species is raised have been used to provide the most extensive evidence for how nutrients can affect the performance of spiders. A number of studies have fed wolf spiders, *Pardosa* spp., with *D. melanogaster* raised on nutrient-poor media with or without dogfood added (Mayntz and Toft, 2000, 2001). *P. amentata* cannot develop to adulthood when fed *D. melanogaster* reared on low-nutrient media and all individuals die within 4 months (Mayntz and Toft, 2001). Addition of dogfood to the rearing media of *D. melanogaster* results in much higher survival and rapid growth of *P. amentata* (i.e. spiders grow from 0.5 to 16 mg in 4 months; Mayntz and Toft, 2001). Dogfood-supplemented flies also increase spider, *Pardosa* spp., performance in a number of other ways. When fed dogfood-supplemented flies, spiders have higher survival when exposed to insecticides (Pedersen *et al.*, 2002) and tolerate toxic prey better (Mayntz and Toft, 2000) than when feeding on low-nutrient flies. These fly manipulations also affect male and female reproduction. Adult males have higher courtship activity and mating success (Lomborg and Toft, 2009) and adult females have faster egg production and greater success at sexual cannibalism (Wilder and Rypstra, 2008b) when fed dogfood-supplemented flies compared to low-nutrient flies. These studies demonstrate that the nutrient content of prey can affect a wide range of performance parameters of spiders. However, it remains unclear which biochemical aspects of prey are affected by adding dogfood to the prey media and is likely a combination of nutrients as spider performance on dogfood-supplemented flies is higher than spider performance when fed flies raised on media supplemented by single nutrient classes (i.e. protein or lipid; Mayntz and Toft, 2001).

4.3 PREY NUTRIENTS

In addition to general manipulations of prey nutrient content, the model system of *D. melanogaster* raised on media of variable composition is able to manipulate more specific aspects of the nutrient content of prey (Mayntz and Toft, 2001). Mayntz and Toft (2001) used this system to test how the addition of macronutrients (i.e. lipid and protein), dietary essential nutrients (i.e. high methionine compared to a more balanced profile of amino acids) and micronutrients (i.e. vitamins) affected the survival and growth of wolf spiders, *P. amentata*. Spider survival was higher than controls when their prey were supplemented with amino acids regardless of whether the supplement was a full range of amino acids or just methionine (Mayntz and Toft, 2001). Spider growth was significantly enhanced by the addition of a full complement of amino acids and lipid (i.e. a range of fatty acids) to the media of the flies but was not affected by the addition of only methionine or vitamins (Mayntz and Toft, 2001). The results of this study suggested that macronutrients such as lipid and protein may be the most important classes of nutrients that vary among prey items and that, in some cases, micronutrients may not be limited or their concentrations may have little effect on spider survival and growth (Mayntz and Toft, 2001).

Following the finding that lipid and protein were important for spider growth, more refined manipulations of these macronutrients were performed to examine if there was an optimal ratio of lipid and protein for spider growth (Jensen *et al.*, 2011a,b; Mayntz and Toft, 2001). By carefully manipulating the quantity of sugar or casein added to *D. melanogaster* media, Jensen *et al.* (2011a) were able to produce six treatments of *D. melanogaster* that spanned a wide range of lipid and protein content. These treatments were used to examine the macronutrient content of prey on which *P. prativaga* and *P. amentata* showed the shortest instar duration and greatest growth (Jensen *et al.*, 2011a). Both species of wolf spiders had a higher growth rate and larger carapace when fed prey with higher protein content, with peak performance when fed flies with a lipid:protein of 0.15–0.25 (Jensen *et al.*, 2011a). *P. amentata* also showed shorter instar durations on high-protein diets, although the same was not true for *P. prativaga* (Jensen *et al.*, 2011a). The final body composition (i.e. amount of lipid and protein in the spider's body) of *P. amentata* was also much more variable than that of *P. prativaga* (Jensen *et al.*, 2011a). The differences in the responses of the congeneric spiders to the nutrient manipulations were likely due to their differences in life history strategies (Jensen *et al.*, 2011a). This study demonstrates that particular concentrations of lipid and protein may support maximal performance in spiders although the results for a given species of spider cannot necessarily be generalized to other related species.

In a complementary study, Jensen *et al.* (2011b) studied prey capture and nutrient extraction of *P. prativaga* feeding on the same six fly treatments. Spiders showed evidence of nutrient regulation at the scale of both killing prey and extracting nutrients from prey. Low levels of prey capture and total energy intake were observed at the same nutrient ratios (~ 0.25 lipid:protein) as peak growth in the prior experiment, suggesting that this may be an optimal ratio of lipid and protein for juvenile spiders (Jensen *et al.*, 2011a,b). When prey with a higher concentration of protein or lipid were provided, *P. prativaga* killed more prey and consumed more total energy (Jensen *et al.*, 2011b). The over-ingestion of high-protein prey allowed spiders to maintain maximum carapace length growth and lean mass growth likely because spiders could catabolize excess protein to compensate for a lack of lipid (Jensen *et al.*, 2011b). However, the over-ingestion of high-lipid flies was ineffective at maintaining lean body growth possibly because spiders became satiated before they could acquire sufficient protein. These results show that spiders are able to balance their intake of lipid and protein in prey and that they likely have an optimum target intake that they are trying to achieve. Further work is needed to examine the nutrient content of potential prey in nature and the consequences of nutrient balancing for prey choice.

In addition to growth, manipulations of protein and lipid in prey have shown that these macronutrients can be important for web building and reproduction. In terms of web building, female *Argiope keyserlingi* had silk decorations in their webs that were twice as long when they were provided with high-protein prey relative to high-lipid prey (Blamires *et al.*, 2009). These decorations are important because they could allow birds to see webs and avoid unintentionally damaging them or increase prey capture at webs (Herberstein *et al.*, 2000). Effects of prey protein content on web building by spiders could be common, given that spider silk is almost entirely composed of amino acids (Foelix, 1996; Nentwig, 1987). While protein is important for growth and web production, lipid may be important for egg production in spiders. The lipid content of spider eggs (25% by dry mass) is similar to the average lipid content of a range of insect prey (24% by dry mass) (Bernard and Allen, 1997; Ramos-Elorduy *et al.*, 1997; Wilder and Rypstra, 2010). Spiders need lipid to directly invest into eggs plus extra energy to fuel the metabolic costs of daily activities and oogenesis. In agreement with the predicted importance of lipid for reproduction, Salomon *et al.* (2008) found that social spider, *Stegodyphus dumicola*, colonies that were supplemented with high-lipid prey had a higher proportion of breeding females and females that were heavier than colonies supplemented with high-protein prey. Hence, lipid and protein can have important aspects of a range of performance measures in spiders. Further research is needed to examine if there are optimum ingestion rates of lipid and protein and if these rates differ among performance measures (e.g. growth, reproduction and survival) as has been observed in other animals (Lee *et al.*, 2008; Maklakov *et al.*, 2008).

4.4 CONCLUSIONS: WHAT DO SPIDERS NEED?

One strength of the literature on spider nutrition is that much of this research has been done on one group, the wolf spiders (Lycosidae), and one genus within this group, *Pardosa* (Table 2). This has allowed for more specific study and a greater depth of information on how a wide range of biochemical aspects of prey (e.g. toxins, prey species, general nutrients, nutrient classes) can affect a wide range of behavioural, physiological and life history parameters (Table 2). However, the focus of research on *Pardosa* is also a limitation of the literature on spider nutrition because very little is known about the nutritional ecology of other spiders and the generality of conclusions based on studies of *Pardosa* spp. for other spiders. There are likely to be significant differences in the nutritional ecology of different groups of spiders, given differences in hunting and life history strategies. For example, there are even differences in how two species of *Pardosa* respond to the same nutrient manipulations (Jensen *et al.*, 2011a). Comparative and phylogenetic studies of spider nutrition using species that differ in particular ecological, morphological, physiological or life history traits will be critical for understanding which aspects of spider nutrition vary depending upon these traits and which are more generalizable.

Another limitation of the literature on spider nutrition is that relatively few studies have examined more detailed biochemical manipulations of nutrients. Most past studies have manipulated prey species or general groups of nutrients and only recently have studies begun manipulating specific nutrient classes (e.g. lipids and protein). Yet, authors have long speculated that specific dietary essential nutrients could limit the performance of spiders (e.g. amino acids, Greenstone, 1979; fatty acids like linoleic acid, Mayntz and Toft, 2001; Uetz *et al.*, 1992). It is time to move beyond studies that ask questions such as ''Does nutrition affect spider performance?'' and begin asking more specific questions related to how specific macronutrients affect the performance and ecology of spiders. Studies that manipulate specific nutrients have the simultaneous benefit of addressing whether or not nutrition is important and beginning to unravel which nutrients are important. Even studies that are not explicitly interested in nutritional physiology (e.g. studies that use prey nutrients to manipulate the condition of a spider for studies of sexual selection) should focus on specific manipulations of nutrients, especially ecologically relevant nutrients that are known to vary in food items in nature. While studying smaller scales of nutrients may be more difficult, it may provide exciting and unexpected results (e.g. effects of methionine versus other amino acids on *D. melanogaster* longevity and reproduction; Grandison *et al.*, 2009).

Finally, while research on spider nutrition has spanned several scales (e.g. prey species, general nutrients and specific nutrient classes), there has been little integration among these scales. Integration can occur either from larger to smaller scales (e.g. How do prey species affect spider performance and what are the specific nutrients responsible for these effects?) or vice versa (e.g. What nutrients are important for spider performance and how do these nutrients vary in prey?). Studies of prey toxins have come the closest to integration by examining which prey species have toxins and how those toxins affect spider physiology (Fisker and Toft, 2004; Jespersen and Toft, 2003; Marcussen *et al.*, 1999; Nielsen and Toft, 2000; Oelbermann and Scheu, 2002a; Rickers *et al.*, 2006; Toft and Nielsen, 1997). However, even this collection of studies is not ideal because investigators have yet to identify the biochemical composition of the toxins, which is critical for completing the integration between prey biochemistry, spider physiology and prey species. At the very least, some type of integration between conditions in the field (i.e. more natural) and experimental manipulations in the laboratory (i.e. more rigorously controlled) is valuable because of the complementary nature of these approaches and the importance of data from the field for ensuring that laboratory manipulations are ecologically relevant.

5 Conclusions: Integration and connections

5.1 SCALES OF SPIDER NUTRITION

A range of scales has been studied in spider nutrition including the consumption and utilization of different prey species, general manipulations of nutrients, nutrient classes and nutrient biochemistry (e.g. dietary essential vs. non-essential). Integration among multiple scales may provide the greatest insight into the nutritional requirements and performance of spiders. However, it is often not feasible to integrate all scales in a single study. One of the first difficulties of integration in the study of spider nutrition is determining which scales are most appropriate to examine for the study species. The goal of this section is to describe the various scales at which spider nutrition can be studied as a framework that can be used to help guide the design and analysis of future studies.

Given the potential scales of nutrient limitation in spiders, a hierarchical approach may facilitate an integrative study of the nutritional ecology of spiders. Hierarchical approaches can be taken in the study of nutrient ingestion by spiders and in the study of how nutrients affect spider performance. Such an approach would start at a broad scale and progressively move to smaller scales. Different scales of nutrition may be important for different aspects of spider nutrition including measuring nutrients and the feeding and performance of spiders.

One set of scales that is important for understanding spider nutrition is the scale at which nutrients are measured. These scales range from general (e.g. macronutrient category) to highly specific (e.g. the biochemical structure of

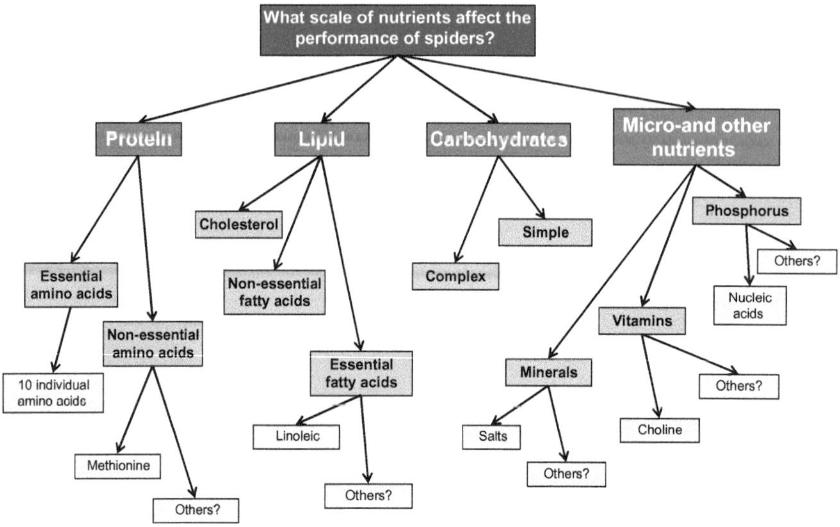

FIG. 1 Diagram representing the different scales at which spider nutrition can be studied. Boxes with darker colours represent higher scales. (For interpretation of the references to colour in this figure legend, the reader is referred to the web version of this book.)

particular amino acids) (Fig. 1). Again, the most efficient way to begin the study of spider nutrition may be to start at higher levels and progressively study lower levels within particularly important categories. The results of studies of nutrients at higher levels can be used to guide where more detailed studies and manipulations of nutrients should be conducted. Placing these studies in an ecological context by integrating laboratory manipulations with field surveys can also help guide studies on the scale of nutrients examined. Surveys of the nutritional content of prey including both the average levels and variation in particular nutrients may help identify which nutrients are the most relevant to manipulate. For example, if arthropod prey consistently have relatively constant and balanced levels of the 10 dietary essential amino acids, then it would be of little relevance to manipulate individual amino acids in this group. However, if two of these amino acids varied significantly among prey items, then studies could conduct focused manipulations on these two specific essential amino acids. Unfortunately, there is so little research on specific aspects of spider nutrition that it is difficult to speculate on what level of nutrient analysis will be most appropriate (e.g. macronutrients vs. dietary essential nutrients). However, carefully planned studies have the potential to rapidly increase our understanding of which scale of nutrients affect spider performance.

Once nutrients are quantified, it is important to understand how spiders acquire those nutrients in nature and the scale of prey nutrients that is most important. A reasonable starting point would be prey taxonomy (order, family, genus or species). For example, different orders of insects likely vary in their gross nutritional composition. Other taxonomic levels of prey organization could also be related to variation in particular aspects of prey nutrition depending upon the scale of nutrients being studied. Below the scale of a species of prey, a range of life history variables could affect the nutrient content of prey including the sex and life history stage of the individual (Wilder *et al.*, 2010). At an even further scale, the recent feeding history of an individual could affect the nutrients they have available in their bodies for spiders to feed upon (Wilder *et al.*, 2010). These levels of analysis could also interact and the characteristics of the specific prey taxa may affect the scale at which generalizations can be made about their nutrient content. For example, there may be large differences between the nutritional content of larval and adult holometabolous insects but much smaller differences in the nutritional content of nymphal and adult hemimetabolous insects (Wilder *et al.*, 2010). It could also be the case that larval holometabolous insects from different orders are more similar to each other in the concentrations of lipid and protein in their body than they are to the adults of their own species (Banjo *et al.*, 2006; Bernard and Allen, 1997; Ramos-Elorduy *et al.*, 1997; Slansky and Scribner, 1985; Wilder and Rypstra, 2010; Wilder *et al.*, 2010). Extensive surveys of the nutritional content of a wide range of insects are needed to examine what scale is most appropriate for categorizing prey and if particular characteristics (e.g. taxonomy, life history stage, life history strategy) are related to the nutritional content of prey for spiders.

Once nutrient intake is quantified or manipulated, it is important to understand the scale at which nutrients affect spider performance. One *a priori* prediction is that more specific scales of nutrients will affect more specific scales of spider performance. For example, overall lipid content of prey might have general effects on lifetime offspring production by females, while specific fatty acids could be responsible for smaller components of offspring production such as the quantity or attractiveness of pheromones used to attract males for mating or the allocation of chemicals to eggs that increase hatching success. The potential congruence between nutrient scale and performance measure scale needs to be tested. Regardless, the scale at which performance is measured should be carefully chosen when examining spider nutrition.

5.2 FRAMEWORKS FOR STUDYING SPIDER NUTRITION

Many studies have provided a solid foundation for the study of spider nutrition. However, many exciting results likely remain uncovered, especially at the scale of macronutrient classes (e.g. protein and lipid) and lower. Lack of rapid progress in understanding spider nutrition may have resulted from several factors. First, there has long been a belief that nutritional content of food is relatively unimportant, or at least much less important, for carnivores compared to other animals and that only the caloric content of food is important for carnivores (Stephens and Krebs, 1986). Recent studies of carnivorous arthropods, including studies of spiders, have begun to change this view and demonstrate that carnivores can be selective of their nutrient intake and can be strongly affected by the nutrient content of their diet (Table 2; e.g. Mayntz and Toft, 2001; Mayntz *et al.*, 2005; Raubenheimer *et al.*, 2007). In addition, methodological constraints on the ability to easily and accurately measure and manipulate specific nutrients have also slowed progress on spider nutrition. Recent advances have made it much easier to quantify nutrient content in prey (e.g. simple techniques like chloroform lipid extraction and protein assays; Horowitz, 2002; Mayntz *et al.*, 2005; Wilder *et al.*, 2010), quantify spider ingestion (e.g. Mayntz *et al.*, 2005; Wilder *et al.*, 2010) and manipulate specific nutrient content of prey (e.g. Jensen *et al.*, 2011a,b; Mayntz and Toft, 2001).

Another factor that may have impeded a more explicit study of spider nutrition is a general lack of *a priori* hypotheses or predictions about how specific nutrients may affect spider performance. Studies that test specific hypotheses often have a greater impact and connect better with more general theories than studies that test for general differences. General frameworks may help in creating *a priori* hypotheses and predictions and may also help guide research in a way that allows smaller studies to provide insight into larger questions.

One framework that may be helpful for forming *a priori* hypotheses about the effects of specific nutrients on spiders is a life history framework (Fig. 2; Boggs, 2009). The nutritional requirements of spiders likely depend upon the predominant behavioural and physiological processes in which they are engaged and consideration of these processes can help for *a priori* hypotheses about nutritional requirements (Boggs, 2009). For example, juvenile spiders need to grow and a major component of growth is the metabolic requirements of synthesizing new tissue and the nutrients that are invested in the structure of new tissue. Hence, spider growth might be expected to be maximized on diets with relatively higher protein content, given the amount of protein that may be needed to build new tissue (Jensen *et al.*, 2011a). This is in contrast to adult male spiders, which have stopped growing and predominantly spend time searching for and courting females (Kotiaho *et al.*, 1998; Watson and Lighton, 1994). Adult males likely have relatively lower requirements for protein (e.g. to produce small quantities of sperm) and a much higher requirement for lipid (although protein can and does provide some energy for spider metabolism, lipid is likely a much more effective source of energy, given its high energy density and potential lifespan costs of protein catabolism; Simpson and Raubenheimer, 2009). Predictions can also be made at smaller scales of nutrients. For example, the concentrations of specific fatty acids or other aspects of a female spider's diet might be expected to affect the quantity or composition of pheromones that the female produces (South *et al.*, 2011; Stanley-Samuelson *et al.*, 1988). A life history approach can be used to make specific predictions about how nutrient manipulations affect a spider in a given life stage or for comparisons of spiders among life stages (Boggs, 2009).

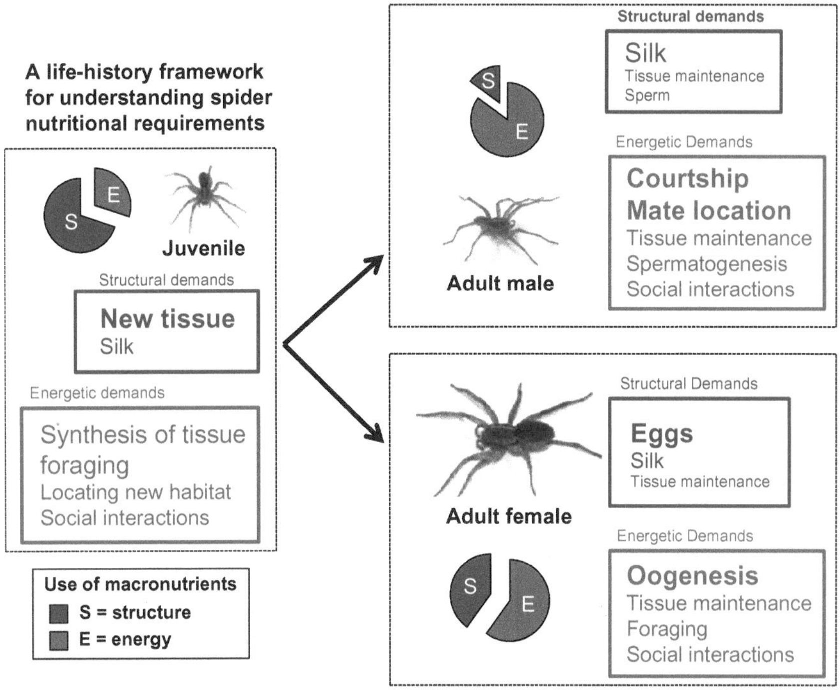

FIG. 2 Hypothesized relationships between life history characteristics and the nutritional requirements of spiders. Major energetic and nutritional demands of the different sexes and life stages are listed with font sizes proportional to the relative requirements of those activities. ''Structural demands'' refer to the use of nutrients to synthesize new molecules in the body, while ''energetic demands'' refer to the catabolism of nutrients to fuel metabolic activities. (See Color Plate 1 at the back of the book.)

Another useful framework is a comparative evolutionary framework for studying spider nutrition. One general comparative hypothesis that can be tested is that spiders, and other carnivorous arthropods, have higher protein requirements in their diet (i.e. fitness is maximized on food with higher concentrations of protein) than herbivores or omnivores. A wide range of other *a priori* hypotheses can be made about specific groups of spiders or comparisons among groups of spiders based on behavioural, physiological or morphological traits or based on their ecology or habitat use. For example, wandering spiders, especially active species, might be expected to have higher energy requirements while web-building spiders might be expected to have higher protein requirements, although this may also vary depending upon whether a species recycles its web or not. Independent evolutionary origins of a range of traits in spiders will facilitate comparative and evolutionary studies of spider nutrition and ecology.

Once *a priori* hypotheses have been developed, there are several frameworks to guide research on nutritional ecology including optimal foraging, classical insect nutrition, ecological stoichiometry and geometric nutrition (Raubenheimer *et al.*, 2009). Frameworks such as geometrical approach may be particularly useful in designing experimental manipulations and understanding the relationships between key nutrients (Simpson and Raubenheimer, 1993, 1995, 1999, 2009; Simpson *et al.*, 2004; Raubenheimer *et al.*, 2009). For example, one tool in the geometrical approach is the use of state-space figures such as "nutrient landscapes" to simultaneously display the quantity of several different nutrients and how they affect performance (Jensen *et al.*, 2011b; Lee *et al.*, 2008; Maklakov *et al.*, 2008; Simpson *et al.*, 2006). However, regardless of the framework used to guide research, it is important to be explicit in the measurement and manipulations of nutrients and to consider the ecological context of the individual species under study (e.g. potential prey, life history, phenology) (Raubenheimer *et al.*, 2009).

5.3 CONCLUSIONS

Spider nutrition is exciting for several reasons. Spiders are abundant carnivores in terrestrial ecosystems that can have large effects on ecosystem and community dynamics (Hawlena and Schmitz, 2010; Schmitz, 2003; Schmitz *et al.*, 2010). Research on spider nutrition has potential to provide insight into the mechanisms through which spiders affect populations, communities and ecosystems including prey choice, intraguild predation, numerical and functional responses, and the transfer and deposition of nutrients (Denno and Fagan, 2003; Fagan and Denno, 2004; Fagan *et al.*, 2002; Raubenheimer *et al.*, 2009). Spiders may also represent a model system for the study of carnivore biology, given their abundance, ease of use in both the laboratory and the field and diversity of behaviour, physiology, morphology and ecology (Barth, 1985; Foelix, 1996; Nentwig, 1987; Ubick *et al.*, 2005; Wise, 1993). Better integration between the field and the laboratory, including ecologically relevant manipulations of the quantities and ratios of nutrients in prey, is critical for advancing our understanding of spider nutrition. In addition, greater integration among disciplines (e.g. molecular biology, physiology, ecology and evolution) will be critical for advancing the understanding of spider nutrition and has the potential to provide exciting insight into the ecology and evolution of spiders and other carnivores.

Acknowledgements

Many thanks to Jerome Casas and Stephen Simpson for the opportunity to write this review and to two anonymous reviewers for helpful comments. I am indebted to a number of colleagues and mentors who have been instrumental in teaching me about spiders, nutrition and biology, in general, especially Micky Eubanks, David Mayntz, Douglas Meikle, Ann Rypstra, Stephen Simpson and Søren Toft.

References

Agusti, N., Shayler, S. P., Harwood, J. D., Vaughan, I. P., Sunderland, K. D. and Symondson, W. O. C. (2003). Collembola as alternative prey sustaining spiders in arable ecosystems: prey detection within predators using molecular markers. *Mol. Ecol.* **12**, 3467–3475.

Andersen, S. O. (1970). Amino acid composition of spider silks. *Comp. Biochem. Physiol.* **35**, 705–711.

Andersen, S. O. (1979). Biochemistry of the insect cuticle. *Annu. Rev. Entomol.* **24**, 29–61.

Andersen, S. O., Hojrup, P. and Roepstorff, P. (1995). Insect cuticular proteins. *Insect Biochem. Mol. Biol.* **25**, 153–176.

Anderson, J. F. (1970). Metabolic rates of spiders. *Comp. Biochem. Physiol.* **33**, 51–72.

Anderson, J. F. (1974). Responses to starvation in the spiders *Lycosa lenta* Hentz and *Filistata hibernalis* (Hentz). *Ecology* **55**, 576–585.

Anderson, J. F. and Prestwich, K. N. (1975). The fluid pressure pumps of spiders (Chelicerata, Araneae). *Z. Morph.* **81**, 257–277.

Anderson, J. F. and Prestwich, K. N. (1982). Respiratory gas exchange in spiders. *Physiol. Zool.* **55**, 72–90.

Anderson, C. M. and Tillinghast, E. K. (1980). GABA and taurine derivatives on the adhesive spiral of the orb webs of *Argiope* spiders, and their possible behavioural significance. *Physiol. Entomol.* **5**, 101–106.

Banjo, A. D., Lawal, O. A. and Songonuga, E. A. (2006). The nutritional value of fourteen species of edible insects in southwestern Nigeria. *Afr. J. Biotechnol.* **5**, 298–301.

Barth, F. G. (1985). Neurobiology of Arachnids. Springer-Verlag, New York.

Beckerman, A., Petchey, O. L. and Morin, P. J. (2010). Adaptive foragers and community ecology: linking individuals to communities and ecosystems. *Funct. Ecol.* **24**, 1–6.

Behmer, S. T. (2009). Insect herbivore nutrient regulation. *Annu. Rev. Entomol.* **54**, 165–187.

Bernard, J. B. and Allen, M. E. (1997). Feeding captive insectivorous animals: nutritional aspects of insects as food. Nutrition Advisory Group, American Zoo and Aquarium Association.

Bilde, T. and Toft, S. (1998). Quantifying food limitation of arthropod predators in the field. *Oecologia* **115**, 54–58.

Bilde, T. and Toft, S. (2000). Evaluation of prey for the spider *Dicymbium brevisetosum* Locket (Araneae: Linyphiidae) in single-species and mixed-species diets. *Ekologia* **19**, 9–18.

Blackledge, T. A., Scharff, N., Coddington, J. A., Szuts, T., Wenzel, J. W., Hayashi, C. Y. and Agnarsson, I. (2009). Reconstructing web evolution and spider diversification in the molecular era. *Proc. Natl. Acad. Sci. USA* **106**, 5229–5234.

Blamires, S. J. (2010). Nutritional implications for sexual cannibalism in a sexually dimorphic orb web spider. *Austral Ecol.* **36**, 389–394.

Blamires, S. J., Hochuli, D. F. and Thompson, M. B. (2009). Prey protein influences growth and decoration building in the orb web spider *Argiope keyserlingi*. *Ecol. Entomol.* **34**, 545–550.

Boggs, C. L. (2009). Understanding insect life histories and senescence through a resource allocation lens. *Funct. Ecol.* **23**, 27–37.
Breed, A. L., Levine, V. D., Peakall, D. B. and Witt, P. N. (1964). The fate of intact orb web of the spider *Araneus diadematus*. *Behaviour* **23**, 43–60.
Brooks, R. C., Simpson, S. J. and Raubenheimer, D. (2010). The price of protein: combining evolutionary and economic analysis to understand excessive energy consumption. *Obes. Rev.* **11**, 887–894.
Buddington, R. K., Chen, J. W. and Diamond, J. M. (1991). Dietary regulation of intestinal brush-border sugar and amino acid transport in carnivores. *Am. J. Physiol* **261**, R793–R801.
Carrel, J. E. and Heathcote, R. D. (1976). Heart rate in spiders: influence of body size and foraging energetic. *Science* **193**, 148–150.
Chen, X., Chen, Y., Wu, L., Peng, Y., Chen, J. and Liu, F. (2010). A survey of nectar feeding by spiders in three different habitats. *Bull. Insectol.* **63**, 203–208.
Cherrett, J. M. (1964). The distribution of spiders on the Moor House Nature Reserve. Westmorland. *J. Anim. Ecol.* **33**, 27–48.
Coddington, J. A. (2005). Phylogeny and classification of spiders. In: *Spiders of North America: An Identification Manual* (eds Ubick, D., Paquin, P., Cushing, P. E. and Roth, V.), pp. 18–24. American Arachnological Society.
Cohen, A. C. (1995). Extra-oral digestion in predaceous terrestrial arthropoda. *Annu. Rev. Entomol.* **40**, 85–103.
Craig, C. L., Hsu, M., Kaplan, D. and Pierce, N. E. (1999). A comparison of the silk proteins produced by spiders and insects. *Int. J. Biol. Macromol.* **24**, 109–118.
Craig, C. L., Riekel, C., Herberstein, M. E., Weber, R. S., Kaplan, D. and Pierce, N. E. (2000). Evidence for diet effects on the composition of silk proteins produced by spiders. *Mol. Biol. Evol.* **17**, 1904–1913.
Denne, K. (1982). *Tutelina similis* (Araneae: Salticidae) an ant mimic that feeds on ants. *Am. Arachnol.* **26**, 14.
Denno, R. F. and Fagan, W. F. (2003). Might nitrogen limitation promote omnivory among carnivorous arthropods? *Ecology* **84**, 2522–2541.
Draper, H. H., Philbrick, D. P., Agarwal, S., Meidiger, R. and Phillips, J. P. (2000). Avid uptake of linoleic acid and vitamin E by *Drosophila melanogaster*. *Nutr. Res.* **20**, 113–120.
Eberhard, W. G. (1988). Behavioral flexibility in orb web construction: effects of supplies in different silk glands and spider size and weight. *J. Arachnol.* **16**, 295–302.
Edgar, W. D. (1969). Prey and predators of the wolf spider *Lycosa lugubris*. *J. Zool. Lond.* **159**, 405–411.
Edgar, W. D. (1971). Aspects of the ecological energetics of the wolf spider *Pardosa (Lycosa) lugubris* (Walckenaer). *Oecologia* **7**, 136–154.
Eisert, R. (2011). Hypercarnivory and the brain: protein requirements of cats reconsidered. *J. Comp. Physiol. B* **181**, 1–17.
Elgar, M. A. and Crespi, B. J. (1992). Cannibalism: Ecology and Evolution among Diverse Taxa. Oxford University Press, New York.
Elser, J. J., Sterner, R. W., Gorokhova, E., Fagan, W. F., Markow, T. A., Cotner, J. B., Harrison, J. F., Hobbie, S. E., Odell, G. M. and Weider, L. J. (2000). Biological stoichiometry from genes to ecosystems. *Ecol. Lett.* **3**, 540–550.
Elser, J. J., Watts, T., Bitler, B. and Markow, T. A. (2006). Ontogenetic coupling of growth rate with RNA and P contents in five species of *Drosophila*. *Funct. Ecol.* **20**, 846–856.
Fagan, W. F. and Denno, R. F. (2004). Stoichiometry of actual vs. potential predator-prey interactions: insights into nitrogen limitation for arthropod predators. *Ecol. Lett.* **7**, 876–883.

Fagan, W. F., Siemann, E., Mitter, C., Denno, R. F., Huberty, A. F., Woods, H. A. and Elser, J. J. (2002). Nitrogen in insects: implications for trophic complexity and species diversification. *Am. Nat.* **160**, 784–802.

Finke, M. D. (2003). Gut loading to enhance the nutrient content of insects as food for reptiles: a mathematical approach. *Zoo Biol.* **22**, 147–162.

Fisker, E. N. and Toft, S. (2004). Effects of chronic exposure to a toxic prey in a generalist predator. *Physiol. Entomol.* **29**, 129–138.

Foelix, R. F. (1996). Biology of Spiders. Oxford University Press, New York.

Ford, M. J. (1977a). Energy costs of the predation strategy of the web-spinning spider *Lepthyphantes zimmermanni* Bertkau (Linyphiidae). *Oecologia* **28**, 341–349.

Ford, M. J. (1977b). Metabolic costs of the predation strategy of the spider *Pardosa amentata* (Clerck) (Lycosidae). *Oecologia* **28**, 333–340.

Fritz, R. S. and Morse, D. H. (1985). Reproductive success and foraging of the crab spider *Misumena vatia*. *Oecologia* **65**, 194–200.

Fromhage, L., Elgar, M. A. and Schneider, J. M. (2005). Faithful without care: the evolution of monogyny. *Evolution* **59**, 1400–1405.

Fu, C., Shao, Z. and Vollrath, F. (2009). Animal silks: their structures. *Chem. Commun.* **43**, 6515–6529.

Furrer, S. and Ward, P. I. (1995). Differential nutrient extraction in the funnel web spider *Agelena labyrinthica*. *Physiol. Entomol.* **20**, 18–22.

Gosline, J. M., DeMont, M. E. and Denny, M. W. (1986). The structure and properties of spider silk. *Endeavor* **10**, 37–43.

Grafen, A. (1989). The phylogenetic regression. *Phil. Trans. R. Soc. Lond. B* **326**, 119–157.

Grandison, R. C., Piper, M. D. W. and Partridge, L. (2009). Amino-acid imbalance explains extension of lifespan by dietary restriction in *Drosophila*. *Nature* **462**, 1061–1064.

Green, A. J. (2001). Mass/length residuals: measures of body condition or generators of spurious results? *Ecology* **82**, 1473–1483.

Greenstone, M. H. (1977). A passive haemagglutination inhibition assay for the identification of stomach contents of invertebrate predators. *J. Appl. Ecol.* **14**, 457–464.

Greenstone, M. H. (1979). Spider feeding behaviour optimizes dietary essential amino acid composition. *Nature* **282**, 501–503.

Greenstone, M. H. and Bennett, A. F. (1980). Foraging strategy and metabolic rate in spiders. *Ecology* **61**, 1255–1259.

Greenstone, M. H. and Shufran, K. A. (2003). Spider predation: species-specific identification of gut contents by polymerase chain reaction. *J. Arachnol.* **31**, 131–134.

Hairston, N. G., Smith, F. E. and Slobodkin, L. B. (1960). Community structure, population control, and competition. *Am. Nat.* **94**, 421–425.

Hardman, J. M. and Turnbull, A. L. (1974). The interaction of spatial heterogeneity, predator competition and the functional response to prey density in a laboratory system of wolf spiders (Araneae: Lycosidae) and fruit flies (Diptera: Drosophilidae). *J. Anim. Ecol.* **43**, 155–171.

Harwood, J. D., Sunderland, K. D. and Symondson, W. O. C. (2004). Prey selection by linyphiid spiders: molecular tracking of the effects of alternate prey on rates of aphid consumption in the field. *Mol. Ecol.* **13**, 3549–3560.

Harwood, J. D., Bostrom, M. R., Hladilek, E. E., Wise, D. H. and Obrycki, J. J. (2007). An order-specific monoclonal antibody to Diptera reveals the impact of alternate prey on spider feeding behavior in a complex food web. *Biol. Control* **41**, 397–407.

Hawlena, D. and Schmitz, O. J. (2010). Herbivore physiological response to predation risk and implications for ecosystem nutrient dynamics. *Proc. Natl. Acad. Sci. USA* **107**, 15503–15507.

Head, G. (1995). Selection on fecundity and variation in the degree of sexual size dimorphism among spider species (Class Araneae). *Evolution* **49**, 776–781.

Heiling, A. M. and Herberstein, M. E. (2000). Interpretations of orb-web variability: a review of past and current ideas. *Ekologia* **19**, 97–106.

Henschel, J. R., Mahsberg, D. and Stumpf, H. (2001). Allochthonous aquatic insects increase predation and decrease herbivory in river shore food webs. *Oikos* **93**, 429–438.

Herberstein, M. E., Craig, C. L., Coddington, J. A. and Elgar, M. A. (2000). The functional significance of silk decorations of orb-web spiders: a critical review of the empirical evidence. *Biol. Rev.* **75**, 649–669.

Hewson-Hughes, A. K., Hewson-Hughes, V. L., Miller, A. T., Hall, S. R., Simpson, S. J. and Raubenheimer, D. (2011). Geometric analysis of macronutrient selection in the adult domestic cat, *Felis catus. J. Exp. Biol.* **214**, 1039–1051.

Higgins, L. and Rankin, M. A. (1999). Nutritional requirements for web synthesis in the tetragnathid spider *Nephila clavipes. Physiol. Entomol.* **24**, 263–270.

Higgins, L. E., Townley, M. A., Tillinghast, E. K. and Rankin, M. A. (2001). Variation in the chemical composition of orb-webs built by the spider *Nephila clavipes. J. Arachnol.* **29**, 82–94.

Hlivko, J. T. and Rypstra, A. L. (2003). Spiders reduce herbivory: nonlethal effects of spiders on the consumption of soybean leaves by beetle pests. *Ann. Entomol. Soc. Am.* **96**, 914–919.

Hoenig, M., Reusch, C. and Peterson, M. E. (2000). Beta cell and insulin antibodies in treated and untreated diabetic cats. *Vet. Immunol. Immunopathol.* **77**, 93–102.

Hormiga, G., Scharff, N. and Coddington, J. A. (2000). The phylogenetic basis of sexual size dimorphism in orb-weaving spiders (Araneae, Orbiculariae). *Syst. Biol.* **49**, 435–462.

Horowitz, H. (2002). *Official Methods of Analysis of the AOAC International.* AOAC Inc., Gaithersburg, MD.

Hurd, L. E. and Eisenberg, R. M. (1990). Arthropod community responses to manipulation of a bitrophic predator guild. *Ecology* **71**, 2107–2114.

Ito, Y. (1964). Preliminary studies on the respiratory energy loss of a spider, *Lycosa pseudoannulata. Res. Popul. Ecol.* **6**, 13–21.

Jackson, R. R. (1992). Eight-legged tricksters. *BioScience* **42**, 590–598.

Jackson, R. R., Pollard, S. D., Nelson, X. J., Edwards, G. B. and Barrion, A. T. (2001). Jumping spiders (Araneae: Salticidae) that feed on nectar. *J. Zool. Lond.* **255**, 25–29.

Jackson, R. R., Nelson, X. J. and Sune, G. O. (2005). A spider that feeds indirectly on vertebrate blood by choosing female mosquitoes as prey. *Proc. Natl. Acad. Sci. USA* **102**, 15155–15160.

Jakob, E. M., Marshall, S. D. and Uetz, G. W. (1996). Estimating fitness: a comparison of body condition indices. *Oikos* **77**, 61–67.

Jensen, K., Mayntz, D., Wang, T., Simpson, S. J. and Overgaard, J. (2010). Metabolic consequences of feeding and fasting on nutritionally different diets in the wolf spider *Pardosa prativaga. J. Insect Physiol.* **56**, 1095–1100.

Jensen, K., Mayntz, D., Toft, S., Raubenheimer, D. and Simpson, S. J. (2011a). Prey nutrient composition has different effects on Pardosa wolf spiders with dissimilar life histories. *Oecologia* **165**, 577–583.

Jensen, K., Mayntz, D., Toft, S., Raubenheimer, D. and Simpson, S. J. (2011b). Nutrient regulation in a predator, the wolf spider *Pardosa prativaga. Anim. Behav.* **81**, 993–999.

Jespersen, L. B. and Toft, S. (2003). Compensatory growth following early nutritional stress in the wolf spider *Pardosa prativaga. Funct. Ecol.* **17**, 737–746.

Kato, C., Iwata, T., Nakano, S. and Kishi, D. (2003). Dynamics of aquatic insect flux affects distribution of riparian web-building spiders. *Oikos* **103**, 113–120.

Kessler, A. (1971). Relation between egg production and food consumption in species of the genus *Pardosa* (Lycosidae, Araneae) under experimental conditions of food-abundance and food-shortage. *Oecologia* **8**, 93–103.

Kiritani, K., Kawahara, S., Sasaba, T. and Nakasuji, F. (1972). Quantitative evaluation of predation by spiders on the green rice leafhopper, *Nephotettix cincticeps* Uhler, by a sight-count method. *Res. Popul. Ecol.* **13**, 187–200.

Kotiaho, J. (1999). Estimating fitness: comparison of body condition indices revisited. *Oikos* **87**, 399–400.

Kotiaho, J. S., Alatalo, R. V., Mappes, J., Nielsen, M. G., Parri, S. and Rivero, A. (1998). Energetic cost of size and sexual signaling in a wolf spider. *Proc. R. Soc. B* **265**, 2203–2209.

Kreiter, N. A. and Wise, D. A. (2001). Prey availability limits fecundity and influences the movement patterns of female fishing spiders. *Oecologia* **127**, 417–424.

Krivan, V. and Schmitz, O. J. (2004). Trait and density mediated indirect interactions in simple food webs. *Oikos* **107**, 239–250.

Lang, A. and Klarenberg, A. J. (1997). Experiments on the foraging behavior of the hunting spider *Pisaura mirabilis* (Araneae: Pisauridae): utilization of single prey items. *Eur. J. Entomol.* **94**, 453–459.

Lease, H. M. and Wolf, B. O. (2011). Lipid content of terrestrial arthropods in relation to body size, phylogeny, ontogeny and sex. *Physiol. Entomol.* **36**, 29–38.

Lee, K. P., Simpson, S. J., Clissold, F. J., Brooks, R., Ballard, J. W. O., Taylor, P. W., Soran, N. and Raubenheimer, D. (2008). Lifespan and reproduction in *Drosophila*: new insights from nutritional geometry. *Proc. Natl. Acad. Sci. USA* **105**, 2498–2503.

Li, D. and Jackson, R. R. (1996). Prey preferences of *Portia fimbriata*, an araneophagic, web-building jumping spider (Araneae: Salticidae) from Queensland. *J. Insect Behav.* **9**, 613–642.

Li, D. and Jackson, R. R. (1997). Influence of diet on survivorship and growth in *Portia fimbriata*, an araneophagic jumping spider (Araneae: Salticidae). *Can. J. Zool.* **75**, 1652–1658.

Linzen, B. and Gallowitz, P. (1975). Enzyme activity patterns in muscles of the lycosid spider, *Cupiennius salei*. *J. Comp. Physiol.* **96**, 101–109.

Lombardi, S. J. and Kaplan, D. L. (1990). The amino acid composition of major ampullate gland silk (dragline) of *Nephila clavipes* (Araneae, Tetragnathidae). *J. Arachnol.* **18**, 297–306.

Lomborg, J. P. and Toft, S. (2009). Nutritional enrichment increases courtship intensity and improves mating success in male spiders. *Behav. Ecol.* **20**, 700–708.

Lowrie, D. C. (1987). Effects of diet on the development of *Loxosceles laeta* (Nicolet) (Araneae, Loxoscelidae). *J. Arachnol.* **15**, 303–308.

Lubin, Y. and Henschel, J. (1996). The influence of food supply on foraging behaviour in a desert spider. *Oecologia* **105**, 64–73.

Lucas, F., Shaw, J. and Smith, S. (1958). The silk fibroins. *Adv. Protein Chem.* **13**, 107–242.

MacDonald, M. L., Rogers, Q. R. and Morris, J. G. (1984). Nutrition of the domestic cat, a mammalian carnivore. *Annu. Rev. Nutr.* **4**, 521–562.

Maklakov, A. A., Simpson, S. J., Zajitschek, F., Hall, M. D., Dessman, J., Clissold, F., Raubenheimer, D., Boduriansky, R. and Brooks, R. C. (2008). Sex-specific fitness effects of nutrient intake on reproduction and lifespan. *Curr. Biol.* **18**, 1062–1066.

Marcussen, B. M., Axelsen, J. A. and Toft, S. (1999). The value of two Collembola species as food for a linyphiid spider. *Entomol. Exp. Appl.* **92**, 29–36.

Markow, T. A., Raphael, B., Dobberfuhl, D., Breitmeyer, C. M., Elser, J. J. and Pfeiler, E. (1999). Elemental stoichiometry of *Drosophila* and their hosts. *Funct. Ecol.* **13**, 78–84.

Matsumura, M., Trafelet-Smith, G. M., Gratton, C., Finke, D. L., Fagan, W. F. and Denno, R. F. (2004). Does intraguild predation enhance predator performance? A stoichiometric perspective. *Ecology* **85**, 2601–2615.

Mayntz, D. and Toft, S. (2000). Effects of nutrient balance on tolerance to low quality prey in a wolf spider (Araneae: Lycosidae). *Ekológia* **3**, 153–158.

Mayntz, D. and Toft, S. (2001). Nutrient composition of the prey's diet affects growth and survivorship of a generalist predator. *Oecologia* **127**, 207–213.

Mayntz, D. and Toft, S. (2006). Nutritional value of cannibalism and the role of starvation and nutrient imbalance for cannibalistic tendencies in a generalist predator. *J. Anim. Ecol.* **75**, 288–297.

Mayntz, D., Toft, S. and Vollrath, F. (2003). Effects of prey quality and availability on the life history of a trap-building predator. *Oikos* **101**, 631–638.

Mayntz, D., Raubenheimer, D., Salomon, M., Toft, S. and Simpson, S. J. (2005). Nutrient-specific foraging in invertebrate predators. *Science* **307**, 111–113.

Mayntz, D., Hunnicke Nielsen, V., Sorenson, A., Toft, S., Raubenheimer, D., Hejlesen, C. and Simpson, S. J. (2009). Balancing of protein and lipid intake by a mammalian carnivore, the mink, *Mustela vison*. *Anim. Behav.* **77**, 349–355.

McGill, B. J., Enquist, B. J., Weiher, E. and Westoby, M. (2006). Rebuilding community ecology from functional traits. *TREE* **21**, 178–185.

Meehan, C. J., Olson, E. J., Reudink, M. W., Kyser, T. K. and Curry, R. L. (2009). Herbivory in a spider through exploitation of an ant-plant mutualism. *Curr. Biol.* **19**, R892–R893.

Miller, J. A. (2007). Repeated evolution of male sacrifice behavior in spiders correlated with genital mutilation. *Evolution* **61**, 1301–1315.

Miyashita, K. (1968). Growth and development of *Lycosa T-insigna* Boes. et Str. (Araneae: Lycosidae) under different feeding conditions. *Appl. Entomol. Zool.* **3**, 81–88.

Miyashita, K. (1969). Effects of locomotor activity, temperature and hunger on the respiratory rate of *Lycosa T-insignata* Boes. et Str. (Araneae: Lycosidae). *Appl. Entomol. Zool.* **4**, 105–113.

Miyashita, T. (1992). Variability in food consumption rate of natural populations in the spider, *Nephila clavata*. *Res. Popul. Ecol.* **34**, 15–28.

Moran, M. D. and Hurd, L. E. (1998). A trophic cascade in a diverse arthropod community caused by a generalist arthropod predator. *Oecologia* **113**, 126–132.

Morse, D. H. (1981). Prey capture by the crab spider *Misumena vatia* (Clerck) (Thomisidae) on three common native flowers. *Am. Midl. Nat.* **105**, 358–367.

Moya-Laraño, J., Orta-Ocaña, J. M., Barrientos, J. A., Bach, C. and Wise, D. H. (2003). Intriguing compensation by adult female spiders for food limitation experienced as juveniles. *Oikos* **101**, 539–548.

Moya-Laraño, J., Vinkovic, D., Allard, C. M. and Foellmer, M. W. (2007). Mass-mediated sex differences in climbing patterns support the gravity hypothesis of sexual size dimorphism. *Web Ecol.* **7**, 106–112.

Moya-Laraño, J., Macías-Ordóñez, R., Blackenhorn, W. U. and Fernández-Montraveta, C. (2008). Analysing body condition: mass, volume or density? *J. Anim. Ecol.* **77**, 1099–1108.

Nelson, X. and Jackson, R. R. (2006). Compound mimicry and trading predators by the males of sexually dimorphic batesian mimics. *Proc. R. Soc. B* **273**, 367–372.

Nentwig, W. (1982). Why do only certain insects escape from a spider's web? *Oecologia* **53**, 412–417.

Nentwig, W. (1983). The prey of web-building spiders compared with feeding experiments (Araneae: Araneidae, Linyphiidae, Pholcidae, Agelenidae). *Oecologia* **56**, 132–139.

Nentwig, W. (1985). Prey analyses of four species of tropical orb-weaving spiders (Araneae: Araneidae) and a comparison with araneids of the temperate zone. *Oecologia* **66**, 580–594.

Nentwig, W. (1986). Non-webbuilding spiders: prey specialists or generalists? *Oecologia* **69**, 571–576.

Nentwig, W. (1987). Ecophysiology of Spiders. Springer-Verlag, New York.

Nielsen, S. A. and Toft, S. (2000). Responses of a detoxification enzyme to diet quality in the wolf spider *Pardosa prativaga*. *Eur. Arachnol.* **19**, 65–70.

Nyffeler, M. (1999). Prey selection of spiders in the field. *J. Arachnol.* **27**, 317–324.

Nyffeler, M. and Benz, G. (1988). Feeding ecology and predatory importance of wolf spiders (*Pardosa* spp.) (Araneae, Lycosidae) in winter wheat fields. *J. Appl. Ent.* **106**, 123–134.

Nyffeler, M. and Breene, R. G. (1990). Evidence of low daily food consumption by wolf spiders in meadowland and comparison with other cursorial hunters. *J. Appl. Entomol.* **110**, 73–81.

Oelbermann, K. and Scheu, S. (2002a). Effects of prey type and mixed diets on survival, growth and development of a generalist predator, *Pardosa lugubris* (Araneae: Lycosidae). *Basic Appl. Ecol.* **3**, 285–291.

Oelbermann, K. and Scheu, S. (2002b). Stable isotope enrichment ($\delta^{15}N$ and $\delta^{13}C$) in a generalist predator (*Pardosa lugubris*, Araneae: Lycosidae): effects of prey quality. *Oecologia* **130**, 337–344.

Olive, C. W. (1982). Behavioral response of a sit-and-wait predator to spatial variation in foraging gain. *Ecology* **63**, 912–920.

Opell, B. D. (1997). The material cost and stickiness of capture threads and the evolution of orb-weaving spiders. *Biol. J. Linn. Soc.* **62**, 443–458.

Opell, B. D. (1998). Economics of spider orb-webs: the benefits of producing adhesive capture thread and of recycling silk. *Funct. Ecol.* **12**, 613–624.

Oxford, G. S. and Gillespie, R. G. (1998). Evolution and ecology of spider coloration. *Annu. Rev. Entomol.* **43**, 619–643.

Parry, D. A. and Brown, R. H. J. (1959). The hydraulic mechanism of the spider leg. *J. Exp. Biol.* **36**, 423–433.

Peakall, D. B. (1966). Regulation of protein production in the silk glands of spiders. *Comp. Biochem. Physiol.* **19**, 253–258.

Peakall, D. B. (1971). Conservation of web proteins in the spider, *Araneus diadematus*. *J. Exp. Zool.* **176**, 257–261.

Peakall, D. B. and Witt, P. N. (1976). The energy budget of an orb web-building spider. *Comp. Biochem. Physiol.* **54**, 187–190.

Pedersen, L. F., Dall, L. G., Sørensen, B. C., Mayntz, D. and Toft, S. (2002). Effects of hunger level and nutrient balance on survival and acetylcholinesterase activity of dimethoate exposed wolf spiders. *Entomol. Exp. Appl.* **103**, 197–204.

Pekár, S. and Toft, S. (2009). Can ant-eating *Zodarion* spiders (Araneae: Zodariidae) develop on a diet optimal for euryphagous arthropod predators? *Physiol. Entomol.* **34**, 195–201.

Pekár, S., Mayntz, D., Ribeiro, T. and Herberstein, M. E. (2010). Specialist ant-eating spiders selectively feed on different body parts to balance nutrient intake. *Anim. Behav.* **79**, 1301–1306.

Philip, B. and Shillington, C. (2010). Effects of prey availability on metabolism and activity in the tarantula *Phormictopus cancerides*. *Can. J. Zool.* **88**, 90–98.

Polis, G. A. and Holt, R. D. (1992). Intraguild predation: the dynamics of complex trophic interactions. *TREE* **7**, 151–154.
Polis, G. A., Myers, C. A. and Holt, R. D. (1989). The ecology and evolution of intraguild predation: potential competitors that eat each other. *Annu. Rev. Ecol. Syst.* **20**, 297–330.
Pollard, S. D., Beck, M. W. and Dodson, G. N. (1995). Why do male crab spiders drink nectar? *Anim. Behav.* **49**, 1443–1448.
Pollierer, M. M., Scheu, S. and Haubert, D. (2010). Taking it to the next level: trophic transfer of marker fatty acids from basal resource to predators. *Soil Biol. Biochem.* **42**, 919–925.
Porter, D. and Vollrath, F. (2009). Silk as a biomimetic ideal for structural polymers. *Adv. Mater.* **21**, 487–492.
Prestwich, K. N. (1977). The energetics of web-building in spiders. *Comp. Biochem. Physiol.* **57**, 321–326.
Prestwich, K. N. (1983a). The roles of aerobic and anaerobic metabolism in active spiders. *Physiol. Zool.* **56**, 122–132.
Prestwich, K. N. (1983b). Anaerobic metabolism in spiders. *Physiol. Zool.* **56**, 112–121.
Prestwich, K. N. (1988). The constraints on maximal activity in spiders. *J. Comp. Physiol. B* **158**, 437–447.
Prestwich, K. N. and Ing, N. H. (1982). The activity of enzymes associated with anaerobic pathways, glycolysis, and the Krebs cycle in spiders. *Comp. Biochem. Physiol.* **72**, 295–302.
Purvis, A. and Rambaut, A. (1995). Comparative analysis by independent contrasts (CAIC): an Apple Macintosh application for analyzing comparative data. *CABIOS* **11**, 247–251.
Ramos-Elorduy, J., Moreno, J. M. P., Prado, E. E., Perez, M. A., Otero, J. L. and de Guevara, O. L. (1997). Nutritional value of edible insects from the State of Oaxaca. *Mexico J. Food Compos. Anal.* **10**, 142–157.
Raubenheimer, D. and Simpson, S. J. (1997). Integrative models of nutrient balancing: application to insects and vertebrates. *Nutr. Res. Rev.* **10**, 151–179.
Raubenheimer, D., Mayntz, D., Simpson, S. J. and Toft, S. (2007). Nutrient-specific compensation following diapause in a predator: implications for intraguild predation. *Ecology* **88**, 2598–2608.
Raubenheimer, D., Simpson, S. J. and Mayntz, D. (2009). Nutrition, ecology and nutritional ecology: towards an integrated framework. *Funct. Ecol.* **23**, 4–16.
Rickers, S., Langel, R. and Scheu, S. (2006). Dietary routing of nutrients from prey to offspring in a generalist predator: effects of prey quality. *Funct. Ecol.* **20**, 124–131.
Riechert, S. E. (1974). Thoughts on the ecological significance of spiders. *Bioscience* **24**, 352–356.
Riechert, S. E. (1988). The energetic costs of fighting. *Am. Zool.* **28**, 877–884.
Riechert, S. E. (1991). Prey abundance vs diet breadth in a spider test system. *Evol. Ecol.* **5**, 327–338.
Riechert, S. E. and Bishop, L. (1990). Prey control by an assemblage of generalist predators: spiders in garden test systems. *Ecology* **71**, 1441–1450.
Riechert, S. E. and Harp, J. M. (1987). Nutritional ecology of spiders. In: *Nutritional Ecology of Insects, Mites and Spiders* (eds Slansky, F. and Rodriguez, J. G.), pp. 645–672. John Wiley & Sons, New York.
Riechert, S. E. and Lockley, T. (1984). Spiders as biological control agents. *Annu. Rev. Entomol.* **29**, 299–320.
Robinson, M. H. and Robinson, B. (1970). Prey caught by a sample population of the spider *Argiope argentata* (Araneae: Araneidae) in Panama: a year's census data. *Zool. J. Linn. Soc.* **49**, 345–358.

Rubenstein, D. I. (1987). Alternative reproductive tactics in the spider *Meta segmentata*. *Behav. Ecol. Sociobiol.* **20**, 229–237.

Rypstra, A. L. (1985). Aggregations of *Nephila clavipes* (L.) (Araneae, Araneidae) in relation to prey availability. *J. Arachnol.* **13**, 71–78.

Rypstra, A. L. and Tirey, R. S. (1991). Prey size, prey perishability and group foraging in a social spiders. *Oecologia* **86**, 25–30.

Salomon, M., Mayntz, D. and Lubin, Y. (2008). Colony nutrition skews reproduction in a social spider. *Behav. Ecol.* **19**, 605–611.

Schmitz, O. J. (1998). Direct and indirect effects of predation and predation risk in old-field interaction webs. *Am. Nat.* **151**, 327–342.

Schmitz, O. J. (2003). Top predator control of plant biodiversity and productivity in an old field ecosystem. *Ecol. Lett.* **6**, 156–163.

Schmitz, O. J. (2004). Trophic cascades: the primacy of trait-mediated indirect interactions. *Ecol. Lett.* **7**, 153–163.

Schmitz, O. J., Hambäck, P. and Beckerman, A. P. (2000). Trophic cascades in terrestrial systems: a review of the effect of top predator removals on plants. *Am. Nat.* **155**, 141–153.

Schmitz, O. J., Hawlena, D. and Trussell, G. C. (2010). Predator control of ecosystem nutrient dynamics. *Ecol. Lett.* **13**, 1199–1209.

Schulte-Hostedde, A. I., Zinner, B., Millar, J. S. and Hickling, G. J. (2005). Restitution of mass-size residuals: validating body condition indices. *Ecology* **86**, 155–163.

Settle, W. H., Ariawan, H., Astuti, E. T., Cahyana, W., Hakim, A. L., Hindayana, D. and Lestari, A. S. (1996). Managing tropical rice pests through conservation of generalist natural enemies and alternative prey. *Ecology* **77**, 1975–1988.

Sherman, P. M. (1994). The orb-web: an energetic and behavioural estimator of a spider's dynamic foraging and reproductive strategies. *Anim. Behav.* **48**, 19–34.

Shreve, S. M., Yi, S. X. and Lee, R. E. (2007). Increased dietary cholesterol enhances cold tolerance in *Drosophila melanogaster*. *Cryoletters* **28**, 33–37.

Sigsgaard, L., Toft, S. and Villareal, S. (2001). Diet-dependent survival, development and fecundity of the spider *Atypena formosana* (Oi) (Araneae: Linyphiidae)—implications for biological control in rice. *Biocontrol Sci. Technol.* **11**, 233–244.

Silva, K. L. and Trigo, J. R. (2002). Structure-activity relationships of pyrrolizidine alkaloids in insect chemical defense against the orb-weaving spider *Nephila clavipes*. *J. Chem. Ecol.* **28**, 657–668.

Simpson, S. J. and Raubenheimer, D. (1993). A multi-level analysis of feeding behaviour: the geometry of nutritional decisions. *Phil. Trans. R. Soc. Lond. B* **342**, 381–402.

Simpson, S. J. and Raubenheimer, D. (1995). The geometric analysis of feeding and nutrition: a user's guide. *J. Insect Physiol.* **41**, 545–553.

Simpson, S. J. and Raubenheimer, D. (1999). Assuaging nutritional complexity: a geometrical approach. *Proc. Nutr. Soc.* **58**, 779–789.

Simpson, S. J. and Raubenheimer, D. (2005). Obesity: the protein leverage hypothesis. *Obes. Rev.* **6**, 133–142.

Simpson, S. J. and Raubenheimer, D. (2009). Macronutrient balance and lifespan. *Aging* **1**, 875–880.

Simpson, S. J., Sibly, R. M., Lee, K. P., Behmer, S. T. and Raubenheimer, D. (2004). Optimal foraging when regulating intake of multiple nutrients. *Anim. Behav.* **68**, 1299–1311.

Simpson, S. J., Sword, G. A., Lorch, P. D. and Couzin, I. D. (2006). Cannibal crickets on a forced march for protein and salt. *Proc. Natl. Acad. Sci. USA* **103**, 4152–4156.

Simpson, S. J., Raubenheimer, D., Charleston, M. A. and Clissold, F. J.the ARC-NZ Vegetation Function Network Herbivory Working Group (2009). Modelling nutritional interactions: from individuals to communities. *TREE* **25**, 53–60.
Slansky, F. Jr. and Rodriguez, J. G. (1987). Nutritional Ecology of Insects, Mites, Spiders, and Related Invertebrates. Wiley, New York.
Slansky, F. Jr. and Scribner, J. M. (1985). Food consumption and utilization. In: *Comprehensive Insect Physiology, Biochemistry and Pharmacology, vol. 4* (eds Kerkut, G. A. and Gilbert, L. I.), pp. 87–163. Pergamon, Oxford.
Smith, R. B. and Mommsen, T. P. (1984). Pollen feeding in an orb-weaving spider. *Science* **226**, 1330–1332.
Snyder, W. E. and Wise, D. H. (2001). Contrasting trophic cascades generated by a community of generalist predators. *Ecology* **82**, 1571–1583.
South, S. H., House, C. M., Moore, A. J., Simpson, S. J. and Hunt, J. (2011). Male cockroaches prefer a high carbohydrate diet that makes them more attractive to females: implications for the study of condition dependence. *Evolution* **65**, 1594–1606.
Spiller, D. A. (1984). Competition between two spider species: experimental field study. *Ecology* **65**, 909–919.
Stanley-Samuelson, D. W., Jurenka, R. A., Cripps, C., Blomquist, G. J. and de Renobales, M. (1988). Fatty acids in insects: composition, metabolism, and biological significance. *Arch. Insect Biochem.* **9**, 1–33.
Stephens, D. W. and Krebs, J. R. (1986). Foraging Theory. Princeton University Press, Princeton, NJ.
Sterner, R. W. and Elser, J. J. (2002). *Ecological Stoichiometry: The Biology of Elements from Molecules to the Biosphere*. Princeton University Press, Princeton, NJ.
Stevenson, R. D. and Woods, W. A. Jr. (2006). Condition indices for conservation: new uses for evolving tools. *Integr. Comp. Biol.* **46**, 1169–1190.
Stoltz, J. A., Hanna, R. and Andrade, M. C. B. (2010). Longevity costs of remaining unmated under dietary restriction. *Funct. Ecol.* **24**, 1270–1280.
Strohmeyer, H. H., Stamp, N. E., Jarzomski, C. M. and Bowers, M. D. (1998). Prey species and prey diet affect growth of invertebrate predators. *Ecol. Entomol.* **23**, 68–79.
Tanaka, K. (1989). Energetic cost of web construction and its effect on web relocation in the web-building spider *Agelena limbata*. *Oecologia* **81**, 459–464.
Tanaka, K. and Ito, Y. (1982). Decrease in respiratory rate in a wolf spider, *Pardosa astrigera*, under starvation. *Res. Popul. Ecol.* **24**, 360–374.
Taylor, R. M. and Bradley, R. A. (2009). Plant nectar increases survival, molting, and foraging in two foliage wandering spiders. *J. Arachnol.* **37**, 232–237.
Taylor, R. M. and Pfannenstiel, R. S. (2008). Nectar feeding by wandering spiders on cotton plants. *Environ. Entomol.* **37**, 996–1002.
Taylor, R. M. and Pfannenstiel, R. S. (2009). How dietary plant nectar affects the survival, growth and fecundity of a cursorial spider *Cheiracanthium inclusum* (Araneae: Miturgidae). *Environ. Entomol.* **38**, 1379–1386.
Tefler, W. H. and Kunkel, J. G. (1991). The function and evolution of insect storage hexamers. *Annu. Rev. Entomol.* **36**, 205–228.
Tillinghast, E. K. and Christenson, T. (1984). Observations on the chemical composition of the web of *Nephila clavipes* (Araneae, Araneidae). *J. Arachnol.* **12**, 69–74.
Toft, S. (1995). Value of the aphid *Rhopalosiphum padi* as food for cereal spiders. *J. Appl. Ecol.* **32**, 552–560.
Toft, S. (1999). Prey choice and spider fitness. *J. Arachnol.* **27**, 301–307.

Toft, S. and Nielsen, S. A. (1997). Influence of diet quality on the respiratory metabolism of a wolf spider *Pardosa prativaga*. In: *Proceedings of the 16th European Colloquium of Arachnology, Siedlce* (ed Zabka, M.), pp. 301–307.

Toft, S. and Wise, D. H. (1999). Growth, development, and survival of a generalist predator fed single- and mixed-species diets of different quality. *Oecologia* **119**, 191–197.

Toft, S., Li, D. and Mayntz, D. (2010). A specialized araneophagic predator's short-term nutrient utilization depends on the macronutrient content of prey rather than on prey taxonomic affiliation. *Physiol. Entomol.* **35**, 317–327.

Towley, M. A., Bernstein, D. T., Gallagher, K. S. and Tillinghast, E. K. (1991). Comparative study of orb web hygroscopicity and adhesive spider composition in three araneid spiders. *J. Exp. Zool.* **259**, 154–165.

Townley, M. A., Tillinghast, E. K. and Neefus, C. D. (2006). Changes in composition of spider orb web sticky droplets with starvation and web removal, and synthesis of sticky droplet compounds. *J. Exp. Biol.* **209**, 1463–1486.

Tso, I. M. and Severinghaus, L. L. (1998). Silk stealing by *Argyrodes lanyuensis* (Araneae: Theridiidae): a unique form of kleptoparasitism. *Anim. Behav.* **56**, 219–225.

Ubick, D., Paquin, P., Cushing, P. E. and Roth, V. (2005). Spiders of North America: An Identification Manual. American Arachnological Society, 377pp.

Uetz, G. W. (1992). Foraging strategies of spiders. *TREE* **7**, 155–159.

Uetz, G. W., Bischoff, J. and Raver, J. (1992). Survivorship of wolf spiders (Lycosidae) reared on different diets. *J. Arachnol.* **20**, 207–211.

Uetz, G. W., Papke, R. and Kilinc, B. (2002). Influence of feeding regime on body size, body condition and a male secondary sexual character in *Schizocosa ocreata* wolf spiders (Araneae, Lycosidae): condition-dependence in a visual signaling trait. *J. Arachnol.* **30**, 461–469.

Uhl, G., Schmitt, S., Schäfer, M. A. and Blanckenhorn, W. (2004). Food and sex-specific growth strategies in a spider. *Evol. Ecol. Res.* **6**, 523–540.

Van Hook, R. I. (1971). Energy and nutrient dynamics of spider and orthopteran populations in a grassland ecosystem. *Ecol. Monogr.* **41**, 1–26.

Vasconcellos-Neto, J. and Lewinsohn, T. M. (1984). Discrimination and release of unpalatable butterflies by *Nephila clavipes* a neotropical orb-weaving spider. *Ecol. Entomol.* **9**, 337–344.

Venner, S. and Casas, J. (2005). Spider webs designed for rare but life-saving catches. *Proc. R. Soc. B* **272**, 1587–1592.

Vogelei, A. and Greissl, R. (1989). Survival strategies of the crab spider *Thomisus onustus* Walkenaer 1806 (Chelicerata, Arachnida, Thomisidae). *Oecologia* **80**, 513–515.

Vollrath, F. and Selden, P. (2007). The role of behavior in the evolution of spiders, silks and webs. *Annu. Rev. Ecol. Evol. Syst.* **38**, 819–846.

Vollrath, F., Fairbrother, W. J., Williams, R. J. P., Tillinghast, E. K., Berstein, D. T., Gallagher, K. S. and Townley, M. A. (1990). Compounds in the droplets of the orb spider's viscid silk. *Nature* **345**, 526–528.

Waldbauer, G. P. (1968). The consumption and utilization of food. *Adv. Insect Physiol.* **5**, 229–288.

Walker, S. E., Marshall, S. D. and Rypstra, A. L. (1999). The effect of feeding history on retreat construction in the wolf spider *Hogna helluo* (Araneae, Lycosidae). *J. Arachnol.* **27**, 689–691.

Watson, P. J. and Lighton, J. R. B. (1994). Sexual selection and the energetic of copulatory courtship in the Sierra dome spider, *Linyphia litigiosa*. *Anim. Behav.* **48**, 615–626.

White, T. C. R. (1978). The importance of a relative shortage of food in animal ecology. *Oecologia* **33**, 71–86.
Wilder, S. M. and Eubanks, M. D. (2010). Might nitrogen limitation promote omnivory among carnivorous arthropods? Comment. *Ecology* **91**, 3114–3117.
Wilder, S. M. and Rypstra, A. L. (2008a). Sexual size dimorphism mediates the occurrence of state-dependent sexual cannibalism in a wolf spider. *Anim. Behav.* **76**, 447–454.
Wilder, S. M. and Rypstra, A. L. (2008b). Diet quality affects mating behaviour and egg production in a wolf spider. *Anim. Behav.* **76**, 439–445.
Wilder, S. M. and Rypstra, A. L. (2010). Males make poor meals: a comparison of nutrient extraction during sexual cannibalism and predation. *Oecologia* **162**, 617–625.
Wilder, S. M., Mayntz, D., Toft, S., Rypstra, A. L., Pilati, A. and Vanni, M. J. (2010). Intraspecific variation in the quality of prey: a comparison of nutrient presence in prey and nutrient extraction by spiders. *Oikos* **119**, 350–358.
Wilson, R. S. and Bullock, J. (1973). The hydraulic interaction between prosoma and opisthosoma in *Amaurobius ferox*. *Z. Morphol. Tiere* **74**, 221–230.
Wise, D. H. (1975). Food limitation of the spider *Linyphia marginata*: experimental field studies. *Ecology* **56**, 637–646.
Wise, D. H. (1979). Effects of an experimental increase in prey abundance upon the reproductive rates of two orb-weaving spider species (Araneae: Araneidae). *Oecologia* **41**, 289–300.
Wise, D. H. (1983). Competitive mechanisms in a food-limited species: relative importance of interference and exploitative interactions among labyrinth spiders (Araneae: Araneidae). *Oecologia* **58**, 1–9.
Wise, D. H. (1993). Spiders in Ecological Webs. Cambridge University Press, Cambridge, UK.
Wise, D. H. (2006). Cannibalism, food limitation, intraspecific competition, and the regulation of spider populations. *Annu. Rev. Entomol.* **51**, 441–465.
Witt, P. N. (1963). Environment in relation to the behavior of spiders. *Arch. Environ. Health* **7**, 4–12.
Xu, M. and Lewis, R. V. (1990). Structure of a protein superfiber: spider dragline silk. *Proc. Natl. Acad. Sci. USA* **87**, 7120–7124.
Zoran, D. L. (2002). The carnivore connection to nutrition in cats. *JAVMA* **221**, 1559–1567.

Endosymbiont Infections in Spiders

Sara L. Goodacre

School of Biology, University of Nottingham, Nottingham, United Kingdom

Abstract

The maternally transmitted endosymbiotic bacteria *Wolbachia pipientis* and strains of *Rickettsia*, *Spiroplasma* and *Cardinium* have been identified in a wide range of different spiders. One or more endosymbionts have been found within many species and even individual spiders are capable of being infected with more than one microbial type. These bacteria are known to affect the reproduction of other arthropod hosts in a variety of different ways in order to favour their own transmission. In spiders, these types of maternally inherited bacteria could be the mechanism through which traits such as primary sex ratio bias and reproductive incompatibility between populations are achieved.

Of particular note is the finding that endosymbionts can be associated with changes in reproductive and non-reproductive spider behaviour. The peculiarities of the spider reproductive system make spiders a good choice for future studies of the relationship between such maternally acquired infections and more general processes such as sexual selection.

ADVANCES IN INSECT PHYSIOLOGY VOL. 40
ISBN 978-0-12-387668-3
DOI: 10.1016/B978-0-12-387668-3.00003-9

1 Identification of maternally transmitted endosymbiotic bacteria in spiders

The past few years have seen a rapid rise in the study of endosymbiotic bacterial infections in spiders. It is now well established that spiders harbour a range of different endosymbiotic bacteria such as *Wolbachia* (Cordaux *et al.*, 2001; Oh *et al.*, 2000; Rowley *et al.*, 2004), *Rickettsia*, *Spiroplasma* and *Cardinium* (e.g. Duron *et al.*, 2008a,b; Goodacre *et al.*, 2006). Infections have been found in haplogyne spiders (e.g. *Dysdera erythrina* and *Pholcus phalangioides*) and entelegyne spiders (e.g. *Nephila clavata*), suggesting that the division between the two major phylogenetic groupings (haplogyne and entelegyne) poses no fundamental barrier to endosymbiont presence. Instances of multiple infections are observed; for example, *Cardinium* is found to co-occur with *Wolbachia* (Chen *et al.*, 2009; Perlman *et al.*, 2010; Ros and Breeuwer, 2009; Vanthournout *et al.*, 2011). *Cardinium* itself appears to be more widespread in spiders than predicted from its incidence in other arthropod groups (Duron *et al.*, 2008a; Martin and Goodacre, 2009; Perlman *et al.*, 2010; Zchori-Fein and Perlman, 2004).

The endosymbiotic bacteria that infect arthropods belong to different phylogenetic groups and they infect different ranges of host cell. *Wolbachia* and *Rickettsia* are both rod-shaped, non-motile alpha-proteobacteria that belong to the family Rickettsiae. *Cardinium* belongs to a different lineage, the Bacteroidetes, and *Spiroplasmas* belong to yet another bacterial lineage, the Mollicutes. All of these bacterial types are Gram negative (i.e. lacking a cell wall) and are known from studies of other invertebrates to be distributed throughout a variety of host tissue types including digestive tract, reproductive tissue and nervous tissue. *Wolbachia*, *Rickettsia* and *Cardinium* are all found within the cytoplasm of their eukaryotic host cell (Albertson *et al.*, 2009; Kitajima *et al.*, 2007; Werren *et al.*, 2008) but *Spiroplasmas* differ from the other two bacterial types in being found not only within tissues such as the gut but also freely throughout the haemolymph.

The presence of microbes resembling *Rickettsia* within arachnid nervous tissue was noticed as early as 1923 by Cowdry, who documented the presence of *Rickettsia*-like bacteria in the nervous tissue of a jumping spider, *Salticus* spp. and a number of other arachnids. More recent studies of *Rickettsia* infections in ticks have demonstrated that although these bacteria may infect endothelial linings of structures such as the gut in their tick vector, in the host to which they are transferred they also migrate into nervous tissue, causing lesions therein (Joshi and Kovács, 2007). Similar findings of *Wolbachia* have been made in *Drosophila* (Albertson *et al.*, 2009). The involvement of the host's nervous system is of particular interest, given that recent studies in spiders demonstrate a link between endosymbionts and behaviour in two linyphiids, *Erigone atra* and *Pityohyphantes phrygianus* (discussed in more detail in Section 4; Goodacre *et al.*, 2009; Gunnarsson *et al.*, 2009).

Observations of bacteria have also been made from female reproductive tracts of species such as oonopid spiders (Burger *et al.*, 2006) and in the spermatozoa of tetragnathids (Michalik *et al.*, 2006). In the latter case, the overall structure of infected spider sperm did not appear to be aberrant and the reproductive potential of the individual did not appear to be decreased. It has not yet been established whether or not these bacteria are true endosymbionts as opposed to horizontally acquired infectious agents but their presence in reproductive tissue certainly merits that the possibility be considered.

2 Modes of endosymbiont inheritance in spiders and other arachnids

Maternally acquired bacteria are passed directly from mother to her offspring, thus arguments are presented in many cases for natural selection acting on the bacterium rather than the host's genome. Identification of bacteria in spider reproductive tissue (e.g. Burger *et al.*, 2006) and at high concentrations in developing spider embryos (Goodacre, unpublished data) supports this mechanism of inheritance although surveys such as those by Goodacre *et al.* (2006), which identified strains of each of these bacteria in DNA extracted from spider leg tissue, confirm that they are not confined to reproductive tissue types alone.

Vertical transmission of endosymbionts may not be 100% effective, either due to external conditions (termed 'environmental curing') or due to other factors such as the presence of competing bacterial endosymbiont strains. Males have traditionally been thought to be the less likely sex to transmit intracellular bacteria, but the observation that spermatozoa may be closely associated with large numbers of bacteria (Michalik *et al.*, 2006) and the observation that many arthropod species transmit bacteria (including *Rickettsia*) and viruses through their sperm (Afzelius *et al.*, 1989) suggest that this mode of inheritance should not be ruled out.

The evidence for vertical transmission is compelling; however, a consistent lack of congruence between spider host phylogenies and that of the bacterial endosymbionts with which they are infected demonstrates that this is unlikely to be the only mechanism through which they are transmitted, at least on an evolutionary timescale (e.g. Goodacre *et al.*, 2006). Rare instances of horizontal transmission are sufficient to explain discordance between host and bacterial phylogenies. Such events might occur during host feeding on infected prey, as has been shown to occur in *Rickettsia and Spiroplasma* in other arthropods (reviewed by Hurst and Jiggins, 2000) and as has been suggested to occur in spiders by previous authors (Rowley *et al.*, 2004). Horizontal transmission of bacteria could also occur from species with a close physical association with the host such as parasitoids (e.g. Finch, 2005), many of which are infected with endosymbionts. Studies that illustrate this close physical relationship between parasitoids and spiders include that of the wasp *Polysphincta gutfreundi*, which injects the orb-weaving spider *Allocyclosa bifurca* with as-yet unidentified compounds that appear to induce the host to alter its webbing behaviour (Eberhard, 2010).

It has been proposed that reported high incidences of particular bacteria within species, for example, *Cardinium* in *Cybaeus* spiders (Perlman *et al.*, 2010) and *Wolbachia* in *Agelenopsis* species (Baldo *et al.*, 2008) may indicate frequent horizontal transmission. In contrast, the existence of different strains of *Wolbachia* in a woodlouse-eating spider and its prey show that horizontal transmission does not necessarily occur despite regular, close contact (Cordaux *et al.*, 2001). Further, regular horizontal transmission, such as that from infected prey, is expected to generate geographic patterns where sympatric hosts share similar bacterial strains. Evidence so far indicates that this does not appear to be the case. For instance, the Australian *Wolbachia* sequences of Rowley *et al.* (2004) do not form a monophyletic group when analysed together with those of Goodacre *et al.* (2006) and bacteria from British *Tegenaria* group more closely with those from Australian taxa than with bacteria from other British specimens. In summary, it seems unlikely that horizontal transmission of endosymbiotic bacteria in spiders occurs readily on an ecological time frame.

3 Bacterial phenotypes in spiders

W. pipientis, *Rickettsia*, *Spiroplasma* and *Cardinium* are known to affect their hosts in a variety of ways in order to favour their own transmission. There is a large body of literature describing the effect on insect reproduction. Documented effects on insect hosts include distorting the sex ratio (SR) in favour of females or causing cytoplasmic incompatibility (CI) so that uninfected individuals are at a reproductive disadvantage (reviewed by Hurst and Jiggins, 2000). Endosymbiont-driven SR distortion is known to be achieved in other arthropods via a variety of mechanisms such as by feminisation of males, by killing males as embryos or through inducing parthenogenesis. Given sufficient time, responses of the host include the evolution of male killer suppression systems and an 'evolutionary arms race' between host and bacterium ensues (Hornett *et al.*, 2006).

CI in diploid organisms is thought to occur because the DNA of sperm carried by infected males is eliminated from developing zygotes unless rescued by eggs infected with a compatible bacterial strain. Uninfected eggs are incapable of such rescue, male and female chromosomes are unable to combine and embryo mortality occurs. Endosymbiont-driven parthenogenesis in haplo-diploid organisms is thought to be caused by chromosome duplication in unfertilised eggs, resulting in female embryos only. SR distortion through the selective killing of male embryos is shown in at least one case (*Drosophila*) to involve interactions between the bacterium and multiple genes involved in host sex determination. Inactivation of any single gene in the dosage-dependent sex-determining complex prevents male killing from occurring (Veneti *et al.*, 2005).

The precise molecular mechanisms through which processes such as CI and SR distortion are achieved are not completely understood although experimental manipulation shows the a bacterium can achieve different phenotypes depending on the host that it infects, which implicates both host and bacterial gene involvement in phenotype determination (e.g. Sasaki *et al.*, 2005). There is also evidence to suggest that rapid changes in phenotype can occur, for instance, from CI to male killing (Jaenike, 2007). Studies on ticks indicate that plasmids carried by bacteria have a role in determining phenotype (Baldridge *et al.*, 2010). Plasmid exchange amongst bacteria may indicate one potential way through which rapid change in phenotype is achieved.

Whilst there is a large body of literature describing the effects of endosymbiont infections in groups such as the insects, the bacterial phenotypes of *Wolbachia* and other endosymbionts in spiders have for the most part not been determined. Three exceptions are that of bacterial infections in three species of linyphiid (dwarf) spider. These are *Oedothorax gibbosus*, *P. phrygianus* and *E. atra*. Recent work on *Oedothorax* by Vanthournout *et al.* (2011) presents evidence to show that female-biased SRs are caused by *Wolbachia* infection. Curing females of their infections restores a more even SR although *Wolbachia* infection alone does not appear to explain the observed variation in SR amongst females. Other factors that would be interesting to study in this system include the interactions, if any, between male morphology (there are two distinct male head types), endosymbiont infection and reproductive traits such as mate choice, fecundity and offspring SR.

Work on both *E. atra* and *P. phrygianus* (Goodacre *et al.*, 2009; Gunnarsson *et al.*, 2009) suggest that endosymbiont infections can influence aspects of spider behaviour (this is discussed in further detail in Section 4). In the case of *P. phrygianus*, infection with *Wolbachia* is associated with modification of female post-mating behaviour (Gunnarsson *et al.*, 2009). This behaviour is shown to influence offspring SR (Gunnarsson and Andersson, 1996), and the position adopted by *Wolbachia* infected females is predicted to achieve a greater female bias. Work on *E. atra* also identified that a particular endosymbiont infection (a *Rickettsia*) appears to influence a behaviour not directly linked to reproduction, namely, the tendency to disperse (Goodacre *et al.*, 2009). In this study, PCR-based molecular methods were used to establish the presence of *Rickettsia* in total tissue samples comprising CNS ganglia plus surrounding non-CNS tissue, but it was not possible to show the causal mechanism through which the effect on behaviour was achieved.

Male killing driven by *Wolbachia* has been identified in arachnids such as the pseudoscorpion *Cordylochernes scorpioides* (Koop *et al.*, 2009), but *Oedothorax* and *Pityohyphantes* remain, to date, the only spiders in which a direct link between endosymbiont infection and SR bias has been established. There are, however, many other spiders where SR bias is observed that may have a microbial cause (as noted previously by Majerus, 2003 and Goodacre *et al.*, 2006). For instance, SR bias is observed in social spiders such as the theredìid

Anelosimus domingo (Avilés and Maddison, 1991; Avilés *et al.*, 2000), the eresid *Stegodyphus dumicola* (Avilés *et al.*, 1999) and the thomisid *Diaea socialis* (Rowell and Main, 1992). In the case of *A. domingo*, chromosome studies demonstrate that male and female gametes are produced in equal numbers, indicating that a post-zygotic event determines the bias, an observation that could indicate a microbial role. Multiple types of endosymbiotic infection are observed in *Diaea* and *Dysdera* species, where both an absence of males and apparent parthenogenesis are seen in congeneric species (Deeleman-Reinhold, 1986; Rowell and Main, 1992). An apparent absence of males (or a strongly biased SR), which might indicate parthenogenesis, is also noted in *Anapistula caecula* (Baert and Jocque, 1993) and *Hypognatha* species (Levi, 1996).

As early as 1994, it was proposed that parthenogenesis might explain SR bias in spiders (Camacho, 1994) although a microbial mechanism was not then suggested. Parthenogenesis has been observed in species such as the dysderid *Dysdera hungarica* (Deeleman-Reinhold, 1986), the ochyroceratid *Theotima minutissimus* (Edwards *et al.*, 2003) and the amaurobiid *Coelotes* (Shimojana and Nishihira, 2000). However, by itself, parthenogenesis does not necessarily indicate microbial involvement. For instance, the oonopid spider *Triaeris stenaspis*, which undergoes thelytokous parthenogenesis (i.e. producing fertile eggs in the absence of mating), seems to do this not because it is under microbial control but because this reproductive strategy is advantageous under circumstances of population expansion, for instance, during colonisation or invasion of new areas (Korenko *et al.*, 2009).

There is as yet no direct evidence that endosymbionts cause CI or parthenogenesis in spiders although reduced viability of offspring from inter-population crosses, which is an observation consistent with CI, is observed in spiders such as the salticid *Habronattus pugillis* (Masta and Maddison, 2002). There is, however, evidence that endosymbionts can cause both CI and parthenogenesis in other arachnids and the possibility that these also occur in spiders deserves further investigation. For instance, thelytokous parthenogenesis driven by endosymbionts is shown to occur in several species of *Brevipalpus* spider mite (Groot and Breeuwer, 2006) and *Cardinium*-induced CI through female egg mortality is observed in the spider mite *Bryobia sarothamni* (Ros and Breeuwer, 2009). The situation in *Bryobia* is further complicated by the presence of co-infecting *Wolbachia*, which appear to prevent CI from occurring.

4 Effects of endosymbiont infections on spider behaviour

Studies of insects have found associations between endosymbiont infections and behavioural responses in insects. For instance, *Wolbachia* is shown to influence *Drosophila* behavioural response to olfactory cues (Peng *et al.*, 2008) and *Cardinium* is found to influence oviposition behaviour of *Encarsia* parasitoid wasps (Kenyon and Hunter, 2007). As discussed above, the influence of endosymbiont infections on spider hosts remains incompletely understood but there are a few studies that have demonstrated that spider behaviour (mating-specific and non-mating-specific) can be altered by host–microbe interactions. The cases of two linyphiid spiders are discussed below.

4.1 *PITYOHYPHANTES PHRYGIANUS*

The linyphiid spider *P. phrygianus* is unusual amongst members of this group in having a female-biased primary SR. Further, in this species, post-copulatory position of the female is shown to be a predictor of offspring SR (Gunnarsson and Andersson, 1996) and thus there is the potential for females to control the SR of their offspring by altering their post-copulatory position (Gunnarsson *et al.*, 2009). Manipulations of experimentally mated individuals confirm this association although the causal mechanism remains unknown. Sex is determined genetically in *Pityohyphantes* by an XO system and there appears to be no pre-meiotic reason for the absence of males; neither has embryo death, which might be indicative of male killing, been observed. Females have two twisted pairs of spermathecae (Uhl and Gunnarsson, 2001) that are positioned at 90° to one another, in different planes; thus differential storage and use of sperm remains a possible mechanism by which the bias in SR is achieved.

Pityohyphantes carries at least one strain of *Wolbachia* and two strains of *Rickettsia* (Goodacre *et al.*, 2006) with the former at the highest frequency (up to 100% in some populations; Gunnarsson *et al.*, 2009). Recent work indicates that the presence of *Wolbachia* influences the adopted position of females after mating (Gunnarsson *et al.*, 2009). Antibiotic treatment with tetracycline to cure females of their *Wolbachia* infection results in a more even SR, although it was also found to influence the female's post-copulatory position in such a way as is predicted to result in greater numbers of males. It therefore remains unclear whether the effect on SR is achieved directly via an alteration in behaviour or through some independent mechanism. Further experimentation where the position of tetracycline-treated females is manipulated is necessary to disentangle these two factors.

Wolbachia and one of the *Rickettsia* strains were found co-occurring within the same individual; thus there may be further as-yet undocumented interactions between the different types of infection (Goodacre *et al.*, 2006, unpublished data). This could be either through direct contact of bacteria within co-infected individuals or as a result of differences in the bacterial phenotypes that lead them to compete with one another. Results from the study by Gunnarsson *et al.* (2009) and ongoing work suggests that in addition to *P. phrygianus* females controlling SR, male *Wolbachia* infection status may also play a part in influencing mating behaviour in this species (Cotterill *et al.*, unpublished data).

4.2 *ERIGONE ATRA*

E. atra is one of the most frequent long-distance dispersing spiders, using aerial dispersal known as 'ballooning' throughout its life cycle. Ballooning describes dispersal where silk acts as a sail to move the spider passively on air currents. The tendency to disperse in this way is known to have a heritable component (Bonte and Lens, 2007) but also to be determined in part by the environmental conditions that are experienced by juveniles (Bonte *et al.*, 2008). The movement of ballooning spiders once airborne is passive; however, the act of becoming airborne first requires adoption of a specific 'tiptoe' posture, a behaviour that is exclusively associated with this becoming airborne. An effect of *Rickettsia* on adoption of the tiptoeing posture was observed in *E. atra* (Goodacre *et al.*, 2009) with a similar effect observed in the congeneric species *E. dentipalpis* (Goodacre *et al.*, unpublished data). The effect did not appear to be related to reduced overall fitness of *Rickettsia*-infected individuals because other traits such as longevity and fecundity did not appear to be affected.

E. atra females not infected with *Rickettsia* were observed to display tiptoeing behaviour that is a precursor specific to long-distance dispersal more quickly, given the appropriate environmental cues, and to invest a greater proportion of their time in this type of behaviour than infected counterparts. This difference was found both in comparisons between naturally uninfected and infected individuals and in comparisons between individuals that had or had not been cured of their infection using antibiotics. Tests of wild-caught individuals that were caught whilst in the process of long-distance dispersing using aerial traps also confirmed the association between *Rickettsia* infection and dispersal.

The study by Goodacre *et al.* (2009) identified no effect of bacterial endosymbionts on male *E. atra* behaviour despite the fact that both males and females were equally likely to be infected. Effectiveness of antibiotic treatment to remove endosymbiont infections in females was assessed by testing their egg sacs to see if bacteria were being vertically transmitted (antibiotic-treated individuals could not). There was no equivalent test of treatment efficiency in males but the method of treatment used was the same and there was no *a priori* reason to suggest that males were less susceptible to tetracycline. It seems likely, then, that treatment was effective and that the difference in behaviour associated with *Rickettsia* infection is restricted to females despite both sexes carrying the bacterial infection. Such sex-specific behavioural responses are known to occur in other arthropods, for example, *Drosophila*, where male and female brains differ in their responses to the same stimulus (reviewed by Benton, 2010).

Work is currently underway to explore the selective advantages of these types of behavioural change in spider dispersal to either the bacterium or the host (or to both). Theoretical work has shown that selection acting on the host should favour increased male dispersal in endosymbiont-infected meta-populations where male killing occurs because of increased male–male kin competition relative to that of females (Bonte *et al.*, 2009). This situation arises because fewer females produce males, and as a consequence, relatedness amongst surviving males increases. Whilst increased male dispersal may occur and not yet have been detected, this explanation does not fully account for the results of Goodacre *et al.* (2009), who found an absolute decrease in long-distance dispersal tendency of females following antibiotic treatment to cure them of their *Rickettsia* infection. An alternative explanation is that reducing the dispersal tendency of one sex may result in transient biases in local SR within a meta-population structure, which could favour a maternally acquired bacterium (Bonte *et al.*, 2009; Goodacre *et al.*, 2009).

Studies of populations within the United Kingdom show that *E. atra* is also infected with at least one other endosymbiont, *Wolbachia*, which has the potential to influence the reproduction of its host (Goodacre *et al.*, 2009). No significant association was found between levels of *Wolbachia* and *Rickettsia* infection within populations or within individuals, but there is an expectation that selection acting on one bacterial type would also favour co-infecting strains if they are equally likely to be vertically transmitted and are thus in effect in complete linkage disequilibrium. The interactions (if any) between different bacterial variants within *Erigone* are not yet known and warrant further study.

It is possible that reduced dispersal also favours other arthropods with which *E. atra* interacts, such as the parasitoid wasp *Gelis festinans*, which is shown to selectively lay eggs in *E. atra* egg sacs (Baarlen *et al.*, 1996). Reduced host dispersal can lead to increased egg sac density and thus increased parasitism opportunities for the wasp and any endosymbionts that it carries. In this context, it would be interesting to know if *Gelis* wasps are the source of the *Rickettsia* found in spiders, particularly given the opportunity for horizontal transfer, as already discussed. It would also be interesting to test whether or not *Gelis* preferentially parasitise infected or uninfected *Erigone* egg sacs. To date there appear to be no studies addressing these issues.

Regardless of the selective mechanism by which differences in dispersal tendency are achieved, the net result of decreased spider dispersal is reduced opportunities for inter-population genetic exchange (gene flow). This in turn influences the likelihood of significant population differentiation, either due to the fixation of localised adaptations (reduced dispersal has the potential to reduce outbreeding depression) or through the effects of random genetic drift. It also has the potential to limit the rate of spatial population expansion, given that the movement of females capable of laying fertilised eggs is required to establish new populations at the head of an invading front. The potential to limit gene flow directly through an effect on dispersal, rather than indirectly through causing CI, appears not to have been found in any other arthropods prior to the study on *E. atra*. Decreased movement of *Tetranychus* spider mite juveniles through an effect of *Wolbachia* on female ovipositing behaviour is reported (Vala *et al.*, 2004). However, it remains to be determined whether or not the influence of endosymbionts on dispersal is a widespread phenomenon or whether this effect on *E. atra* is an uncommon phenotype.

5 Evolutionary relationships between endosymbionts and their spider hosts

Endosymbiont infections detected in contemporary populations may have been recently acquired, and it is important not to assume that any infection observed currently in nature has had sufficient time to reach equilibrium. This absence of equilibrium applies not only to the bacterium that is manipulating the host but also to maternally inherited host genes (e.g. those on the mitochondrion) that are inherited with the bacterium. A rise in frequency of these genes is expected even if the infection ultimately results in population extinction, which is one of the theoretically possible outcomes, for example, if all males are killed.

In particular cases, however, it is thought that bacterial strains and their hosts have had sufficient time for co-evolution to have occurred to reach a stable equilibrium (reviewed by Charlat *et al.*, 2003). Examples of such co-adaptations that can allow long-term persistence of bacterial infections include the rapid evolution of host genes that suppress male killing (e.g. Hornett *et al.*, 2006). Other examples include that of endosymbiont infections in spider mites, which appear to be harnessed to influence SRs in such a way as favours the host (Vala *et al.*, 2003).

In a similar way, selection for a deviation from a 1:1 SR could be acting to maintain endosymbiont infections in *P. phrygianus* (Gunnarsson *et al.*, 2009). Whilst it is not clear how selection could favour an excess of females in this solitary species where kin selection seems unlikely, theoretically the selective advantage need only be temporary to allow infection frequencies to reach detectable frequencies and the advantage might thus be difficult to detect. A similar argument to explain why species such as *E. atra*, which is often an early coloniser of new areas, could maintain an infection that limits its dispersal tendency is that limited dispersal is advantageous under particular circumstances.

It is also possible that other mutualistic aspects to the spider–bacteria relationship act to maintain infections within a spider population. For instance, Tanaka (1994) and Tanaka and Watanabe (2003) showed that prey consumed by spiders can influence their cold susceptibility if the prey are infected with *Pseudomonas* bacteria. Studies on other arthropods have demonstrated a protective effect of endosymbiotic bacteria against changes in thermal condition (e.g. whitefly temperature sensitivity; Brumin *et al.*, 2011) or of the susceptibility of insects to particular pathogens (Bian *et al.*, 2010; Degnan *et al.*, 2009; Hedges *et al.*, 2008; Moreira *et al.*, 2009).

Early studies on the role of endosymbionts in influencing sensitivity to the common agricultural pesticide deltamethrin appear to show that species such as *E. atra* may have their sensitivity to this agent altered by the presence of *Rickettsia* infection (Sleaford and Goodacre, unpublished data). A similar phenomenon has been observed in whitefly infected with *Rickettsia* (Kontsedalov *et al.*, 2008). In this light is also important to consider that in cases where reproductive success is associated with increased levels of antibacterial compounds in the haemolymph (e.g. *Hygrolycosa* males; Ahtiainen *et al.*, 2006), this success could be because antibacterial compounds are up-regulated in the presence of endosymbionts such as *Wolbachia* rather than being an indicator of fitness *per se*. Although there is evidence that endosymbionts do not always increase an insect's immune response (e.g. Bourtzis *et al.*, 2000; Hurst *et al.*, 2003), the possibility that they stimulate the immune system of spiders differently has not been ruled out.

Multiple infections within species and multiple infections within individuals are not unexpected and have been previously documented in other arthropods (Duron *et al.*, 2010; Majerus *et al.*, 2000; Rousset *et al.*, 1999). Population differentiation itself may facilitate the persistence of individual endosymbionts within different populations whilst the acquisition of superinfections through horizontal transmission may contribute to intra-individual co-infection. Previous studies indicate that multiple infections occur also in spiders (Duron *et al.*, 2008b; Goodacre *et al.*, 2006). The study of the linyphiids *E. atra* (Goodacre *et al.*, 2009), *P. phrygianus* (Gunnarsson *et al.*, 2009) and *O. gibbosus* (Vanthournout *et al.*, 2011) have all demonstrated that a single meta-population can support more than one potentially competing endosymbiont strain.

The presence of competing endosymbiont strains is expected to influence the dynamics of the population as a whole, for example, if neighbouring populations are differently infected with one or more bacterial strains with different phenotypes or if a new strain invades one population. An example of this is observed in *Bryobia* spider mites where *Cardinium*, which causes CI, sometimes co-occurs with *Wolbachia* (Chen *et al.*, 2009; Perlman *et al.*, 2010; Ros and Breeuwer, 2009). No significant association or disassociation has thus far been detected between the two infections but the presence of co-infecting *Wolbachia* appears to prevent CI from occurring even when *Cardinium* is present. Doubly infected *Bryobia* females were more fecund than singly infected individuals, indicating that there are further interactions between the two different bacterial types that may maintain both within a population. Clearly there exists the possibility of selective sweeps on maternally inherited host genes (e.g. those on the mitochondrion), which are expected to rise to high frequency when associated with maternally acquired bacterial strains that are increasing in frequency through manipulating their hosts' reproduction.

6 Are spiders special? Identification of novel bacterial strains in spiders

Thus far the *Wolbachia* and *Cardinium* strains of bacteria identified in spiders appear to belong to clades that infect a much wider range of arthropods (e.g. Goodacre *et al.*, 2006; Rowley *et al.*, 2004). In contrast, most *Rickettsia* from spiders fall into a spider-specific clade, distinct from that containing *Rickettsia* from other arthropods (Goodacre *et al.*, 2006; Vanthournout *et al.*, 2011). There is only one published exception, which is a second strain of *Rickettsia* in the linyphiid *P. phrygianus*, which clusters within the group of *Rickettsia* found in insects. Whilst it is possible that other hosts of the apparently spider-specific *Rickettsia* have not yet been sampled, these data suggest that (1) *Rickettsia* from other arthropod groups do not regularly infect spiders and (2) frequent horizontal transmission of *Rickettsia* between spiders and their insect prey or associated mites is unlikely.

Analysis of the citrate synthase gene sequence indicates a high degree of similarity between one section of the gene in the *Rickettsia* found in spiders and that in closely related bacterial genus, *Bartonella*. The high similarity of this particular region leads to inconsistent placement of the spider *Rickettsia* clade within a wider bacterial phylogeny. Such inconsistencies are suggestive of a known feature of bacterial evolution, namely, genetic exchange through recombination. Bacterial recombination either directly between *Rickettsia* and *Bartonella* or between *Rickettsia* and ancestor predating the *Bartonella/Rickettsia* split could explain the phylogenetic relationships observed (Goodacre *et al.*, 2006).

Differences in resistance and susceptibility to infections between spiders, mites and insects may be a factor determining the non-uniform distribution of bacteria such as *Rickettsia* amongst arthropods. Spiders contain types of antimicrobial peptides not yet found in other arthropod groups, which could be an indicator of one such difference (Lorenzinia *et al.*, 2003). The difference may also account for differences in processes such as microbial colonisation of surface epithelia such as is been suggested by recent work that found novel species of *Friedmaniella* on Japanese spiders and their webs (Iwai *et al.*, 2010).

There is also indirect evidence from the unique bacterial types found in spiders to support the view that some of the novelty might be the result of genetic exchange between spider and bacterial genomes. For instance, genetic exchange between the genomes of the spider host and microbes with which it is infected is inferred from studies of proteins responsible for spider venom toxicity, which indicate that they are very similar to genes found in bacteria (Cordes and Binford, 2003). In cases studied thus far, the proteins in spider and bacterium were so similar that it was not possible to confidently identify which was the donor and which the recipient of the inferred horizontal transfer event.

7 Endosymbiont infections and the evolution of sexually selected traits: spiders as useful model systems

Aspects of spider reproductive biology such as the existence of cryptic female choice and sperm competition have lead to this group been the focus of studies into the importance of sexual selection in diversification and speciation (reviewed by Eberhard, 2004; Gage *et al.*, 2002). Traits that are under sexual selection are those where the advantage conferred on males and females is not the same and a conflict thus arises. One area that has so far received less attention is the relationship between sexually selected traits and maternally inherited infections. The influence of maternally acquired infections on these types of trait is predicted to occur either directly through their effect on SR or through an influence on reproductive behaviour of either sex. Studies illustrating the latter include the finding that endosymbionts alter the oviposition behaviour of *Tetranychus* spider mites (Vala *et al.*, 2004), the male rate of reproduction of isopods (Moreau and Rigaud, 2003) and the lekking behaviour of female *Acraea* butterflies (Jiggins, 2002). There is also theoretical work that indicates that endosymbionts may be maintained in a population through their influence on traits involved in sexual conflict, even in the absence of SR bias or cytoplasmic compatibility (Hayashi *et al.*, 2007).

Studies of endosymbiont infections in spiders indicate that a range of different bacterial types is widespread within the group and that individual species, populations and even individuals may carry more than one bacterial type. The particular strains of some of the bacteria identified are rather different from those in isopods or insects, which have been the subject of most previous studies on sexual selection and endosymbiont infection. Study of spiders thus opens up a potentially new source of bacterial interactions. In some cases, such as the salticid *H. pugillis*, traits consistent with endosymbiont-induced CI have already been observed and attributed to sexual selection because the populations already have divergent male traits (Masta and Maddison, 2002). In this case, CI caused by an endosymbiont is an alternative explanation for the reduced viability of crosses that is observed, although this does not preclude additional effects driven by sexual selection.

Experimental work indicates that spiders are tractable laboratory animals that can be cured of microbial infections with antibiotics, using techniques already established in insects (Goodacre *et al.*, 2009; Gunnarsson *et al.*, 2009). Together with theoretical and comparative studies, manipulations of endosymbiont infections in spiders will allow tests to be made of the relationship between the evolution of host traits under sexual selection, such as cryptic female choice or sperm competition, and maternally acquired infections.

Acknowledgements

I thank the members of the SpiderLab at the University of Nottingham, Jon Bull, Melanie Cotterill, Morito Hayashi, Robert Sleaford and Simon Wright, for stimulating discussions and the School of Biology at the University of Nottingham for providing an excellent environment in which to work. I thank the University of Nottingham and the Research Councils, UK, for support through an RCUK Fellowship and two reviewers for their very helpful comments.

References

Afzelius, B. A., Alberti, G., Dallai, R., Godula, J. and Witalinski, W. (1989). Virus- and *Rickettsia*-infected sperm cells in Arthropods. *J. Invertebr. Pathol.* **53**, 365–377.

Ahtiainen, J. J., Alatalo, R. V., Kortet, R. and Rantal, M. J. (2006). Immune function, dominance and mating success in drumming male wolf spiders *Hygrolycosa rubrofasciata*. *Behav. Ecol. Sociobiol.* **60**, 826–832.

Albertson, R., Casper-Lindley, C., Cao, J., Tram, U. and Sullivan, W. (2009). Symmetric and asymmetric mitotic segregation patterns influence *Wolbachia* distribution in host somatic tissue. *J. Cell Sci.* **122**, 4570–4583.

Avilés, L. and Maddison, W. (1991). When is the sex ratio biased in social spiders?: chromosome studies of embryos and male meiosis in *Anelosimus* species (Araneae, Theridiidae). *J. Arachnol.* **19**, 126–135.

Avilés, L., Varas, C. and Dyreson, E. (1999). Does the African social spider *Stegodyphus dumicola* control the sex of individual offspring? *Behav. Ecol. Sociobiol.* **46**, 237–243.

Avilés, L., McCormack, J., Cutter, A. and Bukowski, T. (2000). Precise, highly female-biased sex ratios in a social spider. *Proc. R. Soc. Lond. B* **267**, 1445–1449.

Baarlen, P., Topping, C. J. and Sunderland, K. D. (1996). Host location by *Gelis festinans*, an eggsac parasitoid of the linyphiid spider *Erigone atra*. *Entomol. Exp. Appl.* **81**, 155–163.

Baert, L. and Jocque, R. (1993). *Anapistula caecula* n. sp., the smallest known female spider (Araneae, Symphytognathidae). *J. Afr. Zool.* **107**, 187–189.

Baldo, L., Ayoub, N. A., Hayashi, C. Y., Russell, J. A., Stahlhut, J. K. and Werren, J. H. (2008). Insight into the routes of *Wolbachia* invasion: high levels of horizontal transfer in the spider genus *Agelenopsis* revealed by *Wolbachia* strain and mitochondrial DNA diversity. *Mol. Ecol.* **17**, 557–569.

Baldridge, G. D., Burkhardt, N. Y., Labruna, M. B., Pacheco, R. C., Paddock, C. D., Williamson, P. C., Billingsley, P. M., Felsheim, R. F., Kurtti, T. J. and Munderloh, U. G. (2010). Wide dispersal and possible multiple origins of low-copy-number plasmids in *Rickettsia* species associated with blood-feeding arthropods. *Appl. Environ. Biol.* **76**, 1718–1731.

Benton, R. (2010). Neuroscience: sexy circuits. *Nature* **468**, 638–640.

Bian, G., Xu, Y., Lu, P., Xie, Y. and Xi, Z. (2010). The endosymbiotic bacterium *Wolbachia* induces resistance to dengue virus in *Aedes aegypti*. *PLoS Pathog.* **6**, e1000833.

Bonte, D. and Lens, L. (2007). Heritability of spider ballooning motivation under different wind velocities. *Evol. Ecol. Res.* **9**, 817–827.

Bonte, D., Travis, J., Swertvaeger, I., De Clercq, N. and Lens, L. (2008). Thermal conditions during juvenile development affect adult dispersal in a spider. *Proc. Natl. Acad. Sci. USA* **105**, 17000–17005.

Bonte, D., Hovestadt, T. and Poethke, H. J. (2009). Sex-specific dispersal and evolutionary rescue in metapopulations infected by male killing endosymbionts. *BMC Evol. Biol.* **9**, 16.

Bourtzis, K., Pettigrew, M. M. and O'Neill, S. L. (2000). *Wolbachia* neither induces nor suppresses transcripts encoding antimicrobial peptides. *Insect Mol. Biol.* **9**, 635–639.

Brumin, M., Kontsedalov, S. and Ghanim, M. (2011). *Rickettsia* influences thermotolerance in the whitefly *Bemisia tabaci* B biotype. *Insect Sci.* **18**, 57–66.

Burger, M., Graber, W., Michalik, P. and Kropf, C. (2006). *Silhouettella loricatula* (Arachnida, Araneae, Oonopidae): a haplogyne spider with complex female genitalia. *J. Morphol.* **267**, 663–677.

Camacho, J. P. M. (1994). Female-biased sex ratio in spiders caused by parthenogenesis? *Hereditas* **120**, 183–185.

Charlat, S., Hurst, G. D. D. and Mercot, H. (2003). Evolutionary consequences of *Wolbachia* infections. *Trends Genet.* **19**, 217–223.

Chen, X. L., Xie, R. R., Li, G. Q. and Hong, X. Y. (2009). Simultaneous detection of endosymbionts *Wolbachia* and *Cardinium* in spider mites (Acari: Tetranychidae) by multiplex PCR. *Int. J.Acarology* **35**, 397–403.

Cordaux, R., Michel-Salzat, A. and Bouchon, D. (2001). *Wolbachia* infection in crustaceans: novel hosts and potential routes for horizontal transmission. *J. Evol. Biol.* **14**, 237–243.

Cordes, M. H. J. and Binford, G. J. (2003). Lateral gene transfer of a dermonecrotic toxin between spiders and bacteria. *Bioinformatics* **22**, 264–268.

Cowdry, E. V. (1923). The distribution if *Rickettsia* in the tissues of insects and arachnids. *J. Exp. Med.* **37**, 431–456(431–457).

Deeleman-Reinhold, C. D. (1986). *Dysdera hungarica* (Kulczynski): a case of parthenogenesis? *Acta. X Int. Arachnol. Cong., Spain* **1**, 25–31.

Degnan, P. H., Yu, Y., Sisneros, N., Wing, R. A. and Moran, N. A. (2009). *Hamiltonella defensa*, genome evolution of protective bacterial endosymbiont from pathogenic ancestors. *Proc. Natl. Acad. Sci. USA* **106**, 9063–9068.

Duron, O., Hurst, G. D. D., Hornett, E. A., Jostling, J. A. and Engelstädter, J. (2008a). High incidence of the maternally inherited bacterium *Cardinium* in spiders. *Mol. Ecol.* **17**, 1427–1437.

Duron, O., Bouchon, D., Boutin, S., Bellamy, L., Zhou, L., Engelstädter, J. and Hurst, G. D. D. (2008b). The diversity of reproductive parasites among arthropods: *Wolbachia* do not walk alone. *BMC Biol.* **6**, 27.

Duron, O., Raymond, M. and Weill, M. (2010). Many compatible *Wolbachia* strains coexist within natural populations of *Culex pipiens* mosquito. *Heredity* **1**, 1–8.

Eberhard, W. G. (2004). Why study spider sex? Special traits of spiders facilitate studies of sperm competition and cryptic female choice. *J. Arachnol.* **32**, 545–556.

Eberhard, W. G. (2010). Recovery of spiders from the effects of parasitic wasps: implications for fine-tuned mechanisms of manipulation. *Anim. Behav.* **79**, 375–383.

Edwards, R. L., Edwards, E. H. and Edwards, A. D. (2003). Observations of *Theotima minutissimus* (Araneae, Ochyroceratidae), a parthenogenetic spider. *J. Arachnol.* **31**, 274–277.

Finch, O. (2005). The parasitoid complex and parasitoid-induced mortality of spiders (Araneae) in a Central European woodland. *J. Nat. Hist.* **39**, 2339–2354.

Gage, M. J. G., Parker, G. A., Nylin, S. and Wiklund, C. (2002). Sexual selection and speciation in mammals, butterflies and spiders. *Proc. R. Soc. Lond. B Biol. Sci.* **269**, 2309–2316.

Goodacre, S. L., Martin, O. Y., Thomas, C. F. G. and Hewitt, G. M. (2006). *Wolbachia* and other endosymbiont infections in spiders. *Mol. Ecol.* **15**, 517–527.

Goodacre, S. L., Martin, O. Y., Bonte, D., Hutchings, L., Woolley, C., Ibrahim, K., Thomas, C. F. G. and Hewitt, G. M. (2009). Microbial modification of host long-distance dispersal capacity. *BMC Biol.* **7**, 32.

Groot, T. and Breeuwer, J. A. J. (2006). *Cardinium* symbionts induce haploid thelytoky in most clones of three closely related *Brevipalpus* species. *Exp. Appl. Acarol.* **39**, 257–271.

Gunnarsson, B. and Andersson, A. (1996). Sex ratio variation in sheet-web spiders: options for female control? *Proc. R. Soc. Lond. B Biol. Sci.* **263**, 1177–1182.

Gunnarsson, B., Goodacre, S. L. and Hewitt, G. M. (2009). Sex ratio, mating behaviour and *Wolbachia* infections in a sheetweb spider. *Biol. J. Linn. Soc.* **98**, 181–186.

Hayashi, T. I., Marshall, J. L. and Gavrilets, S. (2007). The dynamics of sexual conflict over mating rate with endosymbiont infection that affects reproductive phenotypes. *J. Evol. Biol.* **20**, 2154–2164.

Hedges, L. M., Brownlie, J. C., O'Neill, S. L. and Johnson, K. N. (2008). *Wolbachia* and virus protection in insects. *Science* **322**, 702.
Hornett, E. A., Charlat, S., Duplouy, A. M. R., Davies, N., Roderick, G. K., Wedell, N. and Hurst, G. D. D. (2006). Evolution of male-killer suppression in a natural population. *PLoS Biol.* **4**, 1643–1648.
Hurst, G. D. D. and Jiggins, F. M. (2000). Male-killing bacteria in insects: mechanisms, incidence and implications. *Emerg. Infect. Dis.* **6**, 329–336.
Hurst, G. D. D., Anbutsu, H., Kutsukake, M. and Fukatsu, T. (2003). Hidden from the host: *Spiroplasma* bacteria infecting *Drosophila* do not cause an immune response, but are suppressed by ectopic immune activation. *Insect Mol. Biol.* **12**, 93–97.
Iwai, K., Aisaka, K. and Suzuki, M. (2010). *Friedmanniella luteola* sp nov., *Friedmanniella lucida* sp nov., *Friedmanniella okinawensis* sp nov and *Friedmaniella sagamiharensis* sp nov., isolated from spiders. *Int. J. Syst. Evol. Microbiol.* **60**, 113–120.
Jaenike, J. (2007). Spontaneous emergence of a new *Wolbachia* phenotype. *Evolution* **61**, 2244–2252.
Jiggins, F. M. (2002). Widespread 'hilltopping' in *Acraea* butterflies and the origin of sex role-reversed swarming in *Acraea encedon* and *A. encedana*. *Afr. J. Ecol.* **40**, 228–231.
Joshi, S. G. and Kovács, A. D. (2007). *Rickettsia rickettsii* infection causes apoptotic death of cultured cerebellar granule neurons. *J. Med. Microbiol.* **56**, 38–141.
Kenyon, S. G. and Hunter, M. S. (2007). Manipulation of oviposition choice of the parasitoid wasp, *Encarsia pergandiella*, by the endosymbiotic bacterium *Cardinium*. *J. Evol. Biol.* **20**, 707–716.
Kitajima, E. W., Groot, T. V. M., Novelli, V. M., Freitas-Astúa, J., Alberti, G. and de Moraes, G. J. (2007). In situ observation of the *Cardinium* symbionts of *Brevipalpus* (Acari: Tenuipalpidae) by electron microscopy. *Exp. Appl. Acarol.* **42**, 263–271.
Kontsedalov, S., Zchori-Fein, E., Chiel, E., Gottlieb, Y., Inbar, M. and Ghanim, M. (2008). The presence of *Rickettsia* is associated with increased susceptibility of *Bemisia tabaci* (Homoptera: Aleyrodidae) to insecticides. *Pest Manag. Sci.* **64**, 789–792.
Koop, J. L., Zeh, D., Bonilla, M. M. and Zeh, J. (2009). Reproductive compensation favours male-killing *Wolbachia* in a live-bearing host. *Proc. R. Soc. Lond. B Biol. Sci.* **276**, 4021–4028.
Korenko, S., Smerda, J. and Pekar, S. (2009). Life-history of the parthenogenetic oonopid spider, *Triaeris stenaspis* (Araneae: Oonopidae). *Eur. J. Entomol.* **106**, 217–223.
Levi, H. W. (1996). The American orb-weavers *Hypognatha*, *Encyosaccus*, *Xylethrus*, *Gasteracantha*, and *Enacrosoma* (Araneae, Araneidae). *Bull. Museum Comp. Zool. Cambridge* **155**, 89–157.
Lorenzinia, D. M., da Silva, P. I., Fogaçaa, A. C., Buletc, P. and Daffre, S. (2003). Acanthoscurrin: a novel glycine-rich antimicrobial peptide constitutively expressed in the hemocytes of the spider *Acanthoscurria gomesiana*. *Dev. Comp. Immunol.* **27**, 781–791.
Majerus, M. E. N. (2003). Sex Wars: Genes, Bacteria, and Biased Sex Ratios. Princeton University Press, Princeton, NJ.
Majerus, M. E. N., Hinrich, J., Schulenburg, G. V. D. and Zakharov, I. A. (2000). Multiple causes of male-killing in a single sample of the twospot ladybird, *Adalia bipunctata* (Coleoptera: Coccinellidae) from Moscow. *Heredity* **84**, 605–609.
Martin, O. Y. and Goodacre, S. L. (2009). Widespread infections by the bacterial endosymbiont Cardinium in arachnids. *J. Arachnol.* **37**, 106–108.
Masta, S. E. and Maddison, W. P. (2002). Sexual selection driving diversification in jumping spiders. *Proc. R. Soc. Lond. B Biol. Sci.* **99**, 4442–4447.
Michalik, P., Sacher, P. and Alberti, G. (2006). Ultrastructural observations of spermatozoa of several tetragnathid spiders with phylogenetic implications (Araneae, Tetragnathidae). *J. Morphol.* **267**, 129–151.

Moreau, J. and Rigaud, T. (2003). Variable male potential rate of reproduction: high male mating capacity as an adaptation to parasite-induced excess of females? *Proc. R. Soc. Lond. B Biol. Sci.* **270**, 1535–1540.

Moreira, L. A., Iturbe-Ormaetxe, I., Jeffery, J. A., Lu, G. Pyke, A. T., *et al.* (2009). A *Wolbachia* symbiont in *Aedes aegypti* limits infection with Dengue, Chikungunya, and Plasmodium. *Cell* **139**, 1268–1278.

Oh, H. W., Kim, M. I. G., Shin, S. W., Bae, K. S., Ahn, Y. J. and Park, H. Y. (2000). Ultrastructural and molecular identification of a *Wolbachia* endosymbiont in a spider, *Nephila clavata. Insect Mol. Biol.* **9**, 539–543.

Peng, Y., Nielsen, J. E., Cunningham, J. P. and McGraw, E. A. (2008). *Wolbachia* infection alters olfactory-cued locomotion in *Drosophila* spp. *Appl. Environ. Microbiol.* **74**, 3943–3948.

Perlman, S. J., Magnus, S. A. and Copley, C. R. (2010). Pervasive associations between *Cybaeus* spiders and the bacterial symbiont *Cardinium. J. Invertebr. Pathol.* **103**, 150–155.

Ros, V. I. D. and Breeuwer, J. A. J. (2009). The effects of, and interactions between, *Cardinium* and *Wolbachia* in the doubly infected spider mite *Bryobia sarothamni. Heredity* **102**, 413–422.

Rousset, F., Braig, H. R. and O'Neill, S. L. (1999). A stable triple *Wolbachia* infection in *Drosophila* with nearly additive incompatibility effects. *Heredity* **82**, 620–627.

Rowell, D. M. and Main, B. Y. (1992). Sex ratio in the social spider *Diaea socialis* (Araneae, Thomisidae). *J. Arachnol.* **20**, 200–206.

Rowley, S. M., Raven, R. J. and McGraw, E. A. (2004). *Wolbachia pipientis* in Australian spiders. *Curr. Microbiol.* **49**, 208–214.

Sasaki, T., Massaki, N. and Kubo, T. (2005). *Wolbachia* variant that induces two distinct reproductive phenotypes in different hosts. *Heredity* **95**, 389–393.

Shimojana, M. and Nishihira, M. (2000). A new cave-dwelling eyeless spider of the genus *Coelotes* (Araneae: Amaurobiidae) from Okinawa island, the Ryukyu Islands, Japan, with notes on possible parthenogenesis. *Acta Arachnol. Tokyo* **49**, 29–40.

Tanaka, K. (1994). The effect of feeding and gut contents on supercooling in the house spider *Achaearanea tepidariorum* (Araneae, Theridiidae). *Cryo Letters* **15**, 361–366.

Tanaka, K. and Watanabe, M. (2003). Transmission of ice-nucleating active bacteria from a prey reduces cold hardiness of a predator (Araneae: Theridiidae). *Naturwissenschaften* **90**, 449–451.

Uhl, G. and Gunnarsson, B. (2001). Female genitalia in *Pityohyphantes phrygianus*, a spider with a skewed sex ratio. *J. Zool. Lond.* **255**, 367–376.

Vala, F., Van Opijnen, T., Breeuwer, J. A. J. and Sabelis, M. W. (2003). Genetic conflicts over sex ratio: mite-endosymbiont interactions. *Am. Nat.* **161**, 254–266.

Vala, F., Egas, M., Breeuwer, J. A. and Sabelis, M. W. (2004). Wolbachia affects oviposition and mating behaviour of its spider mite host. *J. Evol. Biol.* **17**, 692–700.

Vanthournout, B., Swaegers, J. and Hendrickx, F. (2011). Spiders do not escape reproductive manipulations by *Wolbachia. BMC Evol. Biol.* **11**, 15.

Veneti, Z., Bentley, J. K., Koana, T., Braig, H. R. and Hurst, G. D. D. (2005). A functional dosage compensation complex required for male killing in *Drosophila. Science* **307**, 1461–1463.

Werren, J. H., Baldo, L. and Clark, M. E. (2008). *Wolbachia:* master manipulators of invertebrate biology. *Nat. Rev. Microbiol.* **6**, 741–751.

Zchori-Fein, E. and Perlman, S. J. (2004). Distribution of the bacterial symbiont *Cardinium* in arthropods. *Mol. Ecol.* **13**, 2009–2016.

Grade Changes in Brain–Body Allometry: Morphological and Behavioural Correlates of Brain Size in Miniature Spiders, Insects and Other Invertebrates

William G. Eberhard*,† and William T. Wcislo*

*Smithsonian Tropical Research Institute, Apartado, República de Panamá, USA

†Escuela de Biología, Universidad de Costa Rica, Ciudad Universitaria, Costa Rica

ADVANCES IN INSECT PHYSIOLOGY VOL. 40
ISBN 978-0-12-387668-3
DOI: 10.1016/B978-0-12-387668-3.00004-0

Abstract

We use a recent wave of data to confirm that Haller's rule of brain–body allometry, for smaller species to have relatively larger brains, holds for invertebrates as well as vertebrates. But different invertebrate taxa fall on several different allometric lines (grades). In the smallest animals in several grades, the brain occupies a large fraction (up to approximately 16%) of the total body mass. The brain and the structures enclosing it show morphological alterations suggesting a lack of housing capacity in the head for the brain (e.g. the brain extends into other parts of the body such as the legs or thorax), and other structures normally enclosed in the same area are displaced. Miniature animals may thus sacrifice some morphological aspects of body design to accommodate their disproportionately large CNS. The smallest animals of one such group, orb web spiders, do not show signs of behavioural limitation in web construction compared with larger relatives. We speculate that, because of selection resulting from the high metabolic costs of nervous tissue, grade changes may involve substantial modifications of how brains function, and help explain differences between neuron-profligate vertebrates and invertebrates having far fewer neurons (as few as approximately 200–500 neurons in two groups). Scaling problems associated with small size are of general importance, because many moderate-sized animals have very small free-living immature stages.

> ''. . . one experiences an extraordinary surprise [on comparing the brains of a bee and a dragonfly with those of a fish or an amphibian]. The excellence of the psychic machine does not increase with zoological hierarchy; instead one realizes that in fish and amphibians the nervous centers have undergone an unexpected simplification. Of course their grey matter has increased considerably in mass; but when the structure of their brains is compared with that of bees or dragonflies, they are excessively plain, coarse and rudimentary. It is as if one were to pretend to hold as equals the merits of a rough grandfather clock with the quality of a fine pocket watch, a marvel of fineness, delicacy and precision. As always, in building her marvelous works, nature distinguishes herself much more in her tiny creations than in the large.''
>
> Cajal and Sanchez (1915); transl.

1 Problems of absolute and relative brain size in small animals

Very small animals confront special problems in nervous system design. These problems arise for several reasons: the lower limits of the size at which axons can reliably transmit potentials (Faisal *et al.*, 2005); the relatively large volumes of the nucleus compared with the cytoplasm in very small nerve cells may place lower limits on neuron size (e.g. Grebennikov, 2008; Roth *et al.*, 1994); the general scaling trend known as ''Haller's Rule'' (Rensch, 1948) for smaller animals to have relatively larger brains (e.g. Beutel and Haas, 1998; Beutel *et al.*, 2005;

Bonner, 2006; Polilov, 2008; Seid *et al.*, 2011; Striedter, 2005; Wehner *et al.*, 2007); the possibly severe metabolic costs of building, maintaining and operating relatively large amounts of nervous tissue and neural networks (e.g. Cherniak, 1995; Hasenstaub *et al.*, 2010; Niven and Laughlin, 2008); and the lower size and complexity of dendrites in smaller brains (Wittenberg and Wang, 2007). Haller's Rule is well established for vertebrates (Striedter, 2005), and we show here that it also applies to invertebrates, but its causal basis is still debated (e.g. Gonzalez-Voyer *et al.*, 2009). The metabolic costs of maintaining nerve tissue are especially high, so an automatic consequence of Haller's Rule is that smaller animals pay disproportionately large metabolic costs. Furthermore, the density of metabolic activity in smaller brains is likely to be greater if they have comparable information-processing capabilities, and smaller neurons transmit information more slowly (Niven, 2010; Niven *et al.*, 2007). Within a given taxonomic group, species with smaller brains tend to have smaller numbers of smaller neurons (Niven, 2010).

Size matters, because to a first approximation, an animal's behavioural abilities should correlate positively with the numbers of neurons and their connectivities (see below). The special problems in nervous system design faced by small animals are of general significance, because species with miniature or small-bodied adults are common (e.g. Gaston *et al.*, 2001; Hanken and Wake, 1993), and because many metazoans with large-bodied adults have free-living immatures that are extremely small which have many of the same behavioural needs as adults.

When brain allometries are compared among different taxonomic groups, the allometric line of one group is often displaced upward or downward with respect to the lines of other groups in brain weight versus body weight plots (Fig. 1A) or to the left or right in % brain versus body weight plots (Fig. 1B). These differences are called "grade changes" (Striedter, 2005). For instance, the lines describing the percentage of the body dedicated to the brain versus body weight in amphibians is substantially to the left of that for mammals (Martin, 1981; Striedter, 2005; Fig. 1B).

Here, we discuss the biological significance of these design problems. We survey various arthropods and other invertebrate taxa to empirically demonstrate the constraints that follow from these design problems, and how different taxa may have circumvented them in different ways. Do derived miniature forms conserve comparable behavioural abilities by reducing the size of neurons in order to maintain similar numbers of neurons and their connections? Have they evolved new designs to do more processing with fewer neurons, and so conserve behavioural abilities? Or do smaller animals suffer impaired behavioural performance?

In the size range of insects and spiders, there are reasons to expect that gram for gram the brains of very small species may be functionally inferior. The size of the neuron cell body appears to reach a minimum (about 2 μm in diameter) near the lower end of the range of insect body sizes (body length about 0.3 mm—Grebennikov, 2008), and then does not decrease further in smaller individuals; so very small species probably have reduced numbers of neurons (Beutel *et al.*, 2005, and references therein, Quesada *et al.*, in review), as also

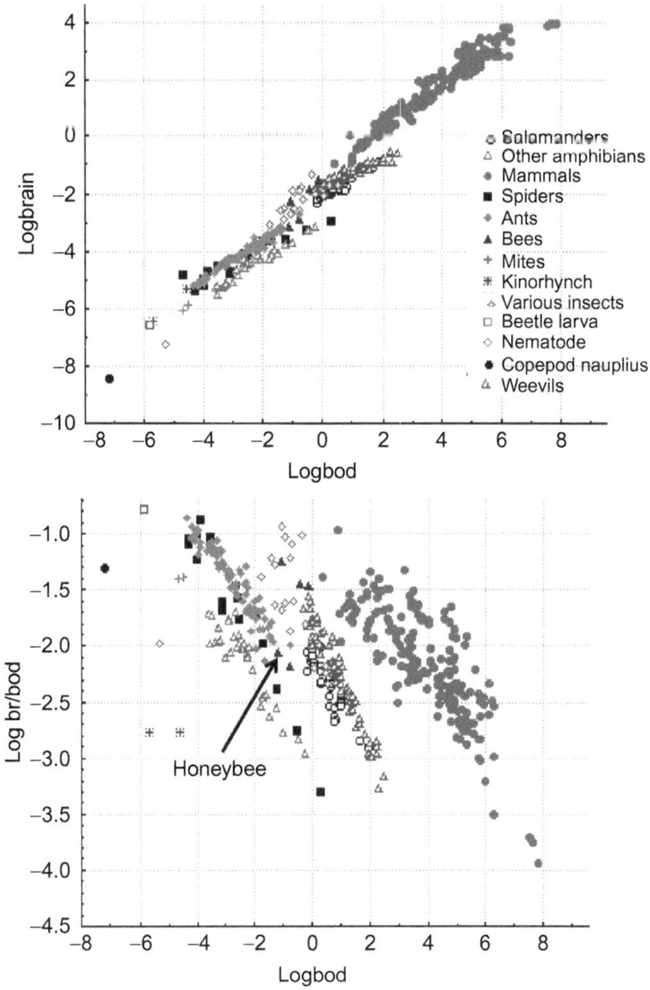

FIG. 1 Brain weight and body weight in vertebrates and assorted invertebrates are given in log–log plots of brain weight versus body weight (A) and of the fraction of the body weight devoted to the brain versus body weight (B) (which permits clearer contrasts between different groups). Sources of data are the following. Vertebrate data were read from graphs in Striedter (2005). Bee weights are from Mares *et al.* (2005), ant weights are from Wehner *et al.* (2007) and Seid *et al.* (2011); weevil weights are from M. Seid, H. Stockwell and W. Wcislo (in preparation). Spider and mite weights were measured directly and from reconstructions of brains from serial sections (Quesada *et al.*, in review, G. Vargas, unpublished). A drawing of a ''typical'' kinorhynch body and brain (Barnes *et al.*, 1993) was deconstructed into simple geometric forms and volumes were estimated based on the cross-sectional body outline in Fig. 296 of Higgins (1983). Two points are given, reflecting uncertainty regarding body weight. Estimated nervous system volume may be underestimated more seriously in this species due to its having a relatively large nerve cord (Bullock and Horridge, 1965). Similar deconstructions and analyses utilized a

occurs in miniature salamanders (Roth *et al.*, 1990). Cells of this size, such as the Kenyon cells of insects, are mostly composed of nucleus, surrounded by only a thin layer of cytoplasm (the nucleus of a neuron of a ptiliid beetle occupies up to 80–90% of the cell—Grebennikov, 2008). Thus a lower limit on cell size may be imposed by nucleus size (Bonner, 2006; Fankhauser *et al.*, 1955), and nucleus size in turn correlates with genome size (Gregory, 2002; Hanken and Wake, 1993; Roth *et al.*, 1988). In addition, there is a lower limit of approximately 0.1 μm on the diameter of functional unmyelinated axons, because stochastic opening of sodium channels in axon membranes can generate action potentials and thus increase noise in very small axons (at least if ion channels are the same in smaller neurons; Faisal *et al.*, 2005). The internal substructures of the brains of very small insects and spiders may also be simpler or fewer in number, though data are scarce (Babu, 1975; Beutel *et al.*, 2005). In absolute terms, smaller insects and spiders tend to have fewer sensory receptors, including chemosensory and tactile setae as well as ommatidia in their compound eyes (Chapman, 1982; Foelix, 1996; Jander and Jander, 2002; Kelber *et al.*, 2006; Mares *et al.*, 2005; Rutowski, 2000; Spaethe and Chittka, 2003; Wcislo, 1995), and their smaller numbers of motor neurons presumably correlate with reduced precision of motor control.

As a consequence of Haller's Rule, miniature animals confront a dilemma. They might maintain information-processing capabilities and behaviours that are comparable to those of larger animals, but either pay the disproportionately high energetic costs (Niven *et al.*, 2007), or evolve more efficient nervous systems that require less tissue to perform comparable tasks (e.g. Wehner, 1987). Alternatively, they could adopt life styles that require reduced behavioural capabilities, which would enable reduced investments in nervous tissue. We describe these alternatives in more detail below (see Section 4).

---

drawing of the anterior end of the nematode *Caenorhabditis elegans* and a photograph of its entire body (by White *et al.*, 1975), and a SEM photo of the entire body plus a cross-section of the body of the first nauplius larva of the copepod *Dactylopusia* sp. (Lacalli, 2009, Figs. 1A and 1D) (this volume may be an underestimate, as the bulbous outline of the nauplius was not obvious in the SEM photo—T. Lacalli, personal communication). The brain volume of this nauplius was calculated by reconstructing sections of the neuropil examined under the TEM (Lacalli, 2009), and an estimate that the cell bodies occupy a volume a bit larger than that of the neuropil (T. Lacalli, personal communication). The body volume of the beetle larva *Mikado* sp. was estimated by deconstruction from Fig. 5 of Polivov and Beutel (2009), assuming a cylindrical cross-section. Their graph (Fig. 20) indicates that the brain occupied about 16.5% of the body volume, so this value was used to estimate the brain volume. All calculations assumed that one cc of brain tissue weighs one gram. All points, other than the copepod nauplius, the beetle larva, and four spiders refer to adults (thus for most groups, additional values for smaller body sizes are lacking). The "brains" of spiders and mites include tissue that in a vertebrate or an insect would be included in the spinal cord or ventral nerve cord, and are thus somewhat higher than they should be to make proper comparisons (for instance, the allometric line for spiders in (b) should probably be below that for ants). (See Color Plate 2 at the back of the book.)

We focus on the relative mass and volumes of the nervous systems rather than the numbers of neurons they contain, because overall size is more directly related to the metabolic costs of a nervous system, thus simplifying deductions with respect to natural selection. Mass is of course only a proxy for energetic cost, because energy consumption is also determined by other factors such as levels of neuron activity and relative amounts of membrane that must be maintained polarized. Pragmatically, data on relative weights are much more common than data on relative numbers of neurons, and data on neuron number frequently lack clarity regarding the methods used to avoid possible biases in stereology (Howard and Reed, 1998). Thus, we analyse alternatives in terms of natural selection to minimize costs, an approach that has provided important insights at the ''macro'' level of morphology (i.e. sizes of different portions of the brain); the organization of behaviour (Clark, 1998); and the ''micro'' level (i.e. distributions of different types of ion gates in neural membranes; see Hasenstaub *et al.*, 2010).

We will assume that to a first approximation that brain size is related to behavioural abilities. There is controversy, however, regarding the relations between brain weight or volume and behavioural capabilities (e.g. Shettleworth, 2010). In a particularly clear statement of the position that the correlation is weak or non-existent, Miklos (1998) surveyed the wide range of numbers of neurons in a variety of organisms (from 302 neurons in a nematode, to 300,000 in a salamander 850,000 in a honey bee, 40 million in the lab mouse, 520 million in an octopus, 85,000 million in a human, and 200,000 million in a whale or an elephant). He concluded that the common belief that ''bigger is better'' (having more neurons enables more complex behaviour) is not supported by facts: ''...no simple relationships exist between gene and neuron number and apparent morphological and behavioral complexity, and the numbers themselves cannot be used in a predictive sense to come to terms with the behavioral complexities of organisms in different evolutionary lineages (p. 203).'' We believe that this conclusion is probably overly pessimistic. In the first place, a lack of a correlation would imply that some animals have excess neural capacity relative to their behavioural needs, and others have insufficient capacity. Such a pattern is unlikely, because the increased cost of maintaining excessive neural capacity is expected to be severely penalized by natural selection (Niven and Laughlin, 2008). In addition, if neural deficiencies lead to behavioural deficiencies, those individuals with insufficient capacity also will be removed by natural selection. Brain-to-body ratios are subject to selection, and can potentially change quickly, as evidenced for example by the fact that the relative brain volumes of many domesticated animal species are relatively smaller than those of their wild-type ancestors (Bernays and Wcislo, 1994; Campi and Krubitzer, 2010), and brains of laboratory-reared guppies are smaller than those of wild guppies (Burns *et al.*, 2009).

Miklos' argument also fails to take into account much data showing that ''more IS better,'' at least with respect to gathering sensory input and information processing (Bonner, 2006; Chittka and Niven, 2009). Absolute size and numbers of sensory receptors, rather than relative size and numbers, set the information-processing capabilities of the nervous system and its efficiency (Herculano-Houzel, 2010). For instance, increased resolution, as well as more subtle or detailed processing of visual information, is clearly related to the numbers of sensory cells in the eye, and the amount of nervous tissue in the portions of the brain that process stimuli from the eye (e.g. Kiltie, 2000; Land and Nilsson, 2002). Compared with diurnal bees, nocturnal bees have relatively larger ommatidia and photoreceptors, which increase the numbers of photons captured to sustain vision at extremely low light levels (Warrant, 2008; Warrant *et al.*, 2004; Wcislo and Tierney, 2009). Similarly, the motor abilities of a structure correlate with the numbers of associated nerves and muscles (Chittka and Niven, 2009). For instance, both the sensory and motor areas of the brain that are devoted to the front paws are greatly enlarged in the dexterous raccoon compared with those in related mammals (Striedter, 2005); the added mobility of the human thumb is associated with extra muscles and presumably extra nerves innervating them, as compared with the thumb of a chimpanzee (Coyne, 2009).

Failures to find correlations between brain and behavioural measurements have probably stemmed from overly broad classifications of behaviour, and overly inclusive measurements of nervous tissue. For instance, the lack of correlation between the total number of neurons in the brains of salamanders and the acrobatic abilities of those living in terrestrial versus arboreal habitats (Miklos, 1998) may have resulted from a lack of focus on the particular portions of the CNS that are involved in this particular behaviour. When questions regarding correlations between behaviour and relative amount of brain tissue are posed in sufficiently specific terms, correlations are often clear (though it is important to remember that correlations do not demonstrate causality in shaping brain evolution—Healy and Rowe, 2007). In pea aphids (*Acyrthosiphon pisum*), for example, there are no overall qualitative differences in brain structure between winged and apterous female morphs, but the winged parthenogenetic females (which have ocelli) have larger central bodies than the apterous females (which lack ocelli; Kollmann *et al.*, 2011); the central body is a brain region thought to be important for visual spatial orientation (Homberg, 2008). Other examples include positive correlations between hovering flight in humming-birds and the relative size of the pretectal nucleus lentiformis mesencephali (Iwaniuk and Wylie, 2007), between the relative size of a visual processing area (superior colliculus) and the degree of dependence on sight in diurnal, nocturnal, and burrowing rodents (Campi and Krubitzer, 2010), and between and the relative size of the hippocampus and the relative importance of spatial memory in the ecology of many (Sherry *et al.*, 1993) but not all (Brodin, 2005) bird species. In a survey of 18 marsupial species, Iwaniuk *et al.* (2000) showed that ability to execute ''finely coordinated forelimb movements'' was not correlated with overall brain size, but rather with the sizes of particular brain regions. In addition, the different expected correlations between volume and behavioural capabilities are likely to be only approximate for other reasons: these include different proportions of glia in different species (they are apparently less abundant in some groups such as nematodes—Bullock and Horridge, 1965), and the need in larger species for longer dendrites and longer and larger diameter axons (to maintain speed of reaction), resulting in a smaller fraction of the brain volume being devoted to information analysis.

Some researchers have proposed that brain size does not correlate with behavioural abilities, but evolves simply through correlated responses to changes in body size (Lande, 1979; Riska and Atchley, 1985). Theoretically we find this argument untenable because it neglects the costly expense of maintaining excess nervous tissue (Hasenstaub *et al.*, 2010; Niven and Laughlin, 2008). And empirically, there are many examples of mosaic evolution of different regions of the brain (Chittka and Niven, 2009). As just mentioned, the claim that there is no evolutionary relationship between brain size and behavioural abilities is probably the result of lumping all CNS tissue into a single category "brain size," and of using such vaguely specified traits as "behaviour" or "intelligence."

One consequence is that the basic tenet of Jerison's (1983) "principle of proper mass"—that the mass of nervous tissue involved in control of a particular function is appropriately sized to the amount of information processing needed for that function—is likely to be at least generally correct (see also Niven, 2010). The fact that brains are highly integrated, so that a given behaviour may involve activity in many different portions, does not detract from this argument, but only makes it more difficult to focus on all of the appropriate portions of the brain. Evolutionary patterns of mosaic brain evolution, which show that different brain regions are differentially enlarged (or reduced) in different lineages (Streidter, 2005), are consistent with the basic assumption that the most useful comparisons for understanding specific behaviours will involve particular brain regions rather than whole brain measurements.

2 Aims of this review

We survey recent evidence from very small insects and spiders regarding their behavioural abilities, and the relative sizes of their brains. We put these morphological data into as large a context as possible, and thus compare insects and spiders with much larger vertebrates, and also with much smaller invertebrates such as kinorhynchs, nematodes and larval crustaceans.

3 Generality of the miniaturization problem

Small-bodied animals are ubiquitous, and in any given habitat, the adults of most animal species are relatively small (e.g. Gaston *et al.*, 2001). In addition, juveniles universally have smaller sizes. Excluding developmental biologists, most scientists working on arthropods concentrate their attention on adults, because they are easier to observe, handle, identify, etc. This bias leads one to forget that most arthropods must function independently just after emerging from the egg, when they are much smaller than an adult. For instance, the adult female of the giant spider *Nephila clavipes* weighs on the order of 2,000 mg, but the nymphs that leave the egg sac and build independent orbs to capture their prey weigh only approximately 0.7 mg (Quesada *et al.*, submitted). For animals in general, biologists tend to think about size-related problems in terms of adult size. Indeed, when discussing "limits to smallness" in the context of body size evolution, Calder (1984, p. 351) wondered whether we are "missing the truth by limiting our concern to *adult* physiology and ecology?" (emphasis added).

Of course there are exceptional species with extensive maternal investment prior to hatching, including tsetse flies, scorpions, cockroaches that nurture their offspring internally (Roth and Willis, 1958), and some others, such as trap-door spiders or solitary bees, that lay very large eggs (Coyle, 1971; Iwata and Sakagami, 1966; Main, 1976). In addition, groups such as many holometabolous insects may have benefited from reducing the behavioural challenges faced by their larvae, since eggs are often deposited in or on a food source, which would shield larvae from possible problems in finding food and defending against predators and parasites. Here the major behavioural problems may involve only relatively simple tasks such as masticating and ingesting food. Even in such cases, larvae may nevertheless actively search their environment, as seen in the *Drosophila* rover morphs (e.g. Osborne *et al.*, 2001). Nothing is known about relative brain size comparing sitter and rover morphs (M. Sokolowski, personal communication). The fact that the environment is less demanding for larvae of holometabolous insects is reflected by the relative numbers of sensory receptors in adults and larvae; after correcting for body size, adult holometabolous insects have far more sensory receptors than do their immature forms; in contrast, numbers are roughly comparable for hemimetabolous species (Bernays and Wcislo, 1994; Chapman, 1982). But even the larvae of tiny ptiliid beetles, which emerge from eggs laid in or on their food have large brains relative to their body size (Polilov and Beutel, 2010), as do many tiny hymenopteran egg parasitoids (Grebennikov, 2008). The important point is that the problems of miniaturization are clearly not limited to species with tiny adults, but are widespread in the animal kingdom.

Hanken and Wake (1993) argued that the historical process of miniaturization involves a phylogenetic hypothesis, and defined miniaturization as the evolution of extremely small body size relative to other species in the same lineage (they referred only to adults). They recognized that "extreme" is a relative term, and that there ". . . is no strict criterion for distinguishing miniaturization from . . . less severe instances of size reduction" (*op cit.* p. 502); thus "miniaturization" occurs at a different absolute body size for beetles than it does for salamanders, for example. Using this definition (which we also use), they showed that miniaturization of adults has evolved independently in numerous living and extinct invertebrates and vertebrates, and in the latter it seems especially common in frogs and fishes. They noted that "...it is difficult to propose any major metazoan taxa that doesn't offer an example" (*op cit.* p. 503). Within arthropods, miniaturization of adults has also evolved independently numerous times (e.g. Beutel *et al.*, 2005; Grebennikov, 2008; Polilov, 2005; Polilov and Beutel, 2010), but it is presently not possible to estimate the actual number because of the lack of lower-level phylogenies in many taxa. Among stingless bees (Meliponini), for example, miniaturization of adults has evolved independently in 11 different genera (Michener, 2001). The importance of phylogenetic information for identifying the directionality of body size change is illustrated by the case of giant orb-weaving spiders *Nephila*, with huge females and small males; phylogenetic studies demonstrate that *Nephila* evolved gigantism in females, rather than dwarfism (miniaturization) in males (Coddington *et al.*, 1997; Hormiga *et al.*, 2000).

The physiological problems associated with small size that were discussed above must be confronted by small animals in general, whether or not their small size is derived. Nevertheless, the most interesting small animals for understanding brain miniaturization are those that arise phylogenetically or ontogenetically from larger animals, and thus have a suite of potential behavioural capacities which they must have either conserve or reduce. In this context, species with relatively small ontogenetic changes in body design and life style (e.g. hemimetabolous insects, most arachnids) are of special interest. In a group without complete metamorphosis, any evolutionary reduction in adult body size is preceded ontogenetically by immature forms that already function at smaller body sizes. Thus the earliest immature stages, soon after they emerge from the egg and begin to fend for themselves, must often constitute the "leading edge" of evolutionary transitions toward smaller body size.

4 Possible solutions to miniaturization problems

We see three extreme options that animals could use to solve the possible brain scaling and behavioural problems that are associated with very small size. These extremes are not mutually exclusive, and different types of adjustments to small size could occur even in a single animal. We first outline these options, and then review the available behavioural and morphological data to evaluate the extent to which different options are exploited by different taxa.

4.1 THE "SIZE LIMITATION" OPTION

One solution to miniaturization problems would be to evolve life styles that are less demanding with respect to behavioural capabilities. This in principle should reduce metabolic costs by reducing the need for expensive brain tissue. Selection should favour reduction in a behavioural capability when the costs associated with this capability outweigh the benefits it confers. The size-limitation hypothesis predicts that very small individuals may be more limited behaviourally, because the greater costs of a CNS that confers equal behavioural capabilities have made it advantageous for these animals to settle for reduced behavioural abilities.

For instance, miniaturization in salamanders has evolved independently 10 times (Hanken and Wake, 1993). In these small salamanders, which depend heavily on vision to capture their prey, the relative sizes of different portions of the brain are altered, and the major visual and motor centres increase in relative size (Roth *et al.*, 1990). Miniaturized salamanders thus appear to have partially redesigned their brains, reducing the relative sizes of some portions. This implies that some as yet unidentified behavioural functions (sensory, analytical, motor?) may have been lost in miniature salamanders, or that they have other compensatory neural mechanisms in these portions of their brains.

4.2 THE "OVER-SIZED BRAIN" OPTION

The selective advantage of particular behavioural capabilities may be great enough to compensate for the relatively high cost of the nervous tissue needed to produce these abilities. Thus smaller animals may evolutionarily opt to maintain these abilities, despite their costs. As a result, the animal will have a relatively large brain.

This hypothesis is at least superficially in accord with the empirically well-established allometric trends that smaller animals generally have relatively larger brains (see Section 1). There is, however, no general consensus regarding the cause(s) of Haller's Rule. Proposed explanations include the scaling relationship between body surface area and volume (Jerison, 1973), and metabolic constraints limiting brain size (Martin, 1981). In *Cataglyphis* ants, colony size correlates positively with larger relative brain size (Wehner *et al.*, 2007), but the generality of this finding is uncertain. There are reasons to expect that the more behaviourally specialized workers in larger social insect colonies will generally have relatively smaller brains when compared with solitary individuals or with less specialized workers in smaller societies (see Gronenberg and Riveros, 2009; Wcislo and Riveros, in preparation).

4.3 THE "ECONOMY OF DESIGN" OPTION

Typically there is massive redundancy in the inputs that animals gather from their environment, and a large discrepancy between the enormous capacities of sensory systems that gather these inputs and the much more limited capacity of the CNS to process them and generate and retain biologically meaningful information (Dusenbery, 1992; Mercer, 1999). This discrepancy can lead to opportunities for removal of redundant information from receptors before it is sent to central processors (Niven, 2010), and for economizing on information processing (Bernays and Wcislo, 1994) by making adjustments that allow smaller numbers of neurons or lower connectivities to produce at least approximately comparable behavioural abilities. Adjustments to improve economy could include a wide variety of possible mechanisms. The list we give below is only preliminary; some adjustments are well documented while others are highly speculative.

The behaviour of some insects (e.g. ants), for instance, is guided by neural mechanisms that are less flexible than those of many vertebrates but that are more efficient in terms of the numbers of neurons involved (the "matched filters" of Wehner, 1987; Collett and Collett, 2002). In insects and spiders, much of the processing of inputs occurs at relatively peripheral neural levels, including both the sensory and the motor processes (Barth, 2002; Chapman, 1982), which offers further opportunities to economize on brain tissue (economies of this sort also occur in vertebrates—e.g. the mechanoreceptors in the skin). One such mechanism is illustrated by the sense organs of some spiders, which arc built of materials whose properties cause them to respond best to relevant stimuli in contexts in which it is biologically advantageous for the spider to respond (Franzl and Barth, 2009). Some of the fundamental spatial aspects of a navigational problem can be incorporated into the spatial design of a sensory receptor (Wehner, 1987), such as the alignment of rhodopsin in microvillar membranes of photoreceptors to detect polarized light (Homberg *et al.*, 2011). Although this may limit the versatility with which input can be handled, it decreases the total input that needs to be centrally processed to generate information. These "tuned" sense organs are constructed so as to screen inputs that would otherwise have to be processed by the CNS (Dusenbery, 1992; Franzl and Barth, 2009).

Another possibility to economize would be to make increased use of graded depolarizations rather than action potentials for transmitting messages, as seems to be especially common in insects (Chittka and Niven, 2009) and nematodes (Bullock and Horridge, 1965). Energetic savings would accrue because analogue signals transmit information more efficiently than digital signals; action potentials require larger ion fluxes (e.g. Clark and Häusser, 2006; Niven *et al.*, 2008; Sarpeshkar, 1998). Another type of efficiency to achieve a behavioural goal involves ordering behavioural sub-routines so that difficult tasks are divided into series of simpler ones. For instance, if a task such as grasping an object is always preceded by preliminary sensory or motor behaviour such as turning to face the object, grasping is much less demanding with respect to sensory and motor control (Clark, 1998). This organization can dramatically increase efficiency by reducing the sensory and motor abilities needed (Clark, 1998; Miller *et al.*, 1960). Miniaturized animals can also realize savings from utilizing the advantageous properties of "small world neural networks" (Watts and Strogatz, 1998), but it is not known if such networks are deployed more extensively in miniature animals relative to large-bodied ones.

At a more microscopic level, savings could come from a variety of adaptations: using muscle plates that extend from muscles toward neurons that allow a single synaptic process to stimulate multiple muscles (Fig. 2; Altun and Hall, 2008; Bullock and Horridge, 1965); using the same neurons for multiple

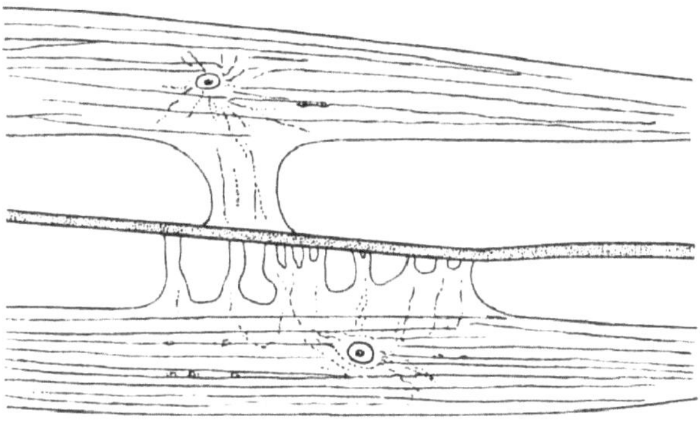

FIG. 2 In nematodes, it is common that a given neuron innervates several muscles, and, as seen in this figure, that muscles send projections to neurons rather than vice versa (Bullock and Horridge, 1965). This could be an efficiency of design that minimizes the numbers and sizes of motor neurons in these tiny animals (after Martini, 1916, from Bullock and Horridge, 1965).

functions; using neuromodulation of networks to produce different outputs from the same neural circuitry (Anderson, 2010; Chittka and Niven, 2009; Katz, 1999); reducing the relative numbers of inter-neurons as opposed to sensory and motor neurons, as is typical in nematodes (Bullock and Horridge, 1965); positioning of neurons and their connections to ''save wire'' and minimize the total length of axons and dendrites (Cherniak, 1995; Cherniak *et al.*, 2004), which is said to be near the theoretical minimum in the nematode, *Caenorhabditis elegans* (Niven, 2010; Fig. 3); or by indirectly controlling cilia through muscles (Keshmirian and Nogrady, 1987). The general importance of minimizing energetic costs is illustrated by the fact that neurons with multiple functions, and neurons that employ analogue signals for transmission over short distances, occur in insects (Bullock and Horridge, 1965) as well as in such ''neuron-profligate'' animals as primates (Niven, 2010; Van Essen and Gallant, 1994). In fact, multiple functionality is probably the ancestral state for neurons (Bullock and Horridge, 1965).

5 Predictions derived from possible solutions to the miniaturization problem

The three options just outlined are expected to be associated with different combinations of behaviour and morphology. These are predictions of possible extreme versions of the options, and actual animals are likely to have mixes of these predicted traits.

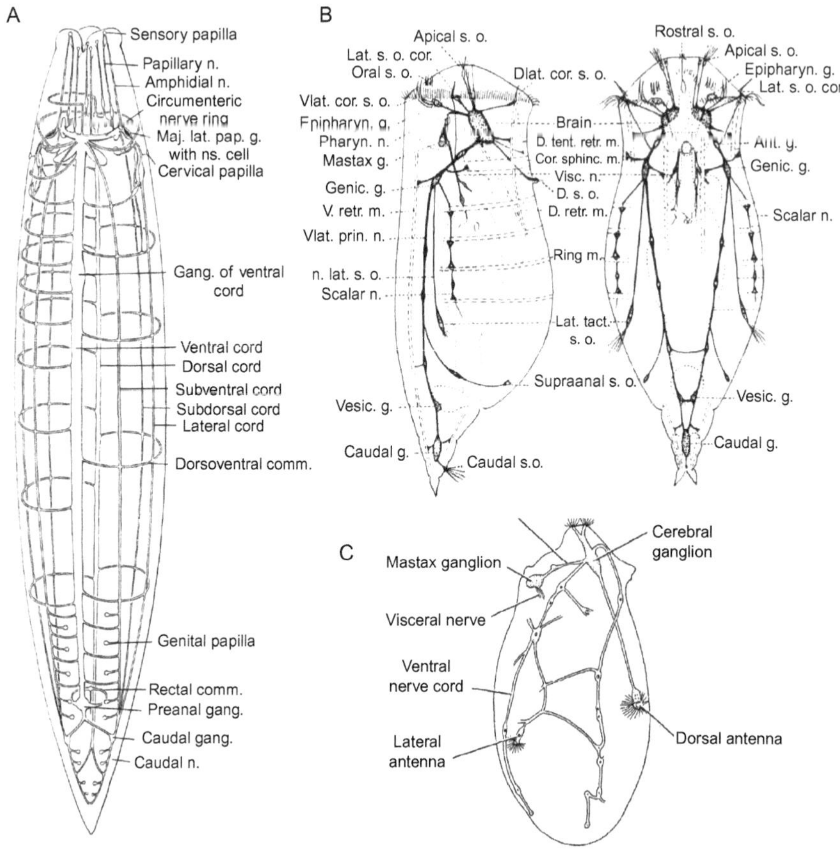

FIG. 3 A possible dimension for efficiency of design is the gross anatomy of nervous systems, illustrated here for two groups of small invertebrates, nematodes (A) and rotifers (B, C). (A) Shows the highly ordered arrangement of the relatively condensed nervous system of the nematode *Ascaris*. In *A. lumbricoides* 63.8% of the animal's total of 254 neurons are in the CNS, and 85.8% of these are connected directly to either muscles or sense organs (Bullock and Horridge, 1965). In another nematode with a somewhat similar design, *C. elegans*, the distribution of neurons brings the length of the connecting processes of neurons close to a theoretical minimum. B and C show the less centralized but nevertheless highly organized nervous systems of two rotifers, a Monogonontan (B) and *Asplanchia* sp. (C). In *A. priodonta*, only about 43% of the approximately 520 neurons are in the brain (Bullock and Horridge, 1965). Can selection for reduced metabolic costs explain these differences in general nervous system designs? (A and B from Bullock and Horridge, 1965; C from Brusca and Brusca, 1990).

5.1 SIZE LIMITATION OPTION

The most directly testable prediction relates to behaviour: smaller organisms should have more limited or impaired behavioural capacities. Morphologically, the hypothesis predicts that smaller animals will not have "special" (derived) morphological compensations in the sizes and shapes of their brains or of the parts of their bodies that house them.

5.2 OVER-SIZED BRAIN OPTION

The behavioural prediction of this hypothesis is that smaller animals will not have impaired behavioural capacities. On the morphological side, smaller animals should have relatively large brains. These could, at least in some cases, either extend into additional areas of the body, or be housed in areas that are expanded or deformed to increase the volume available for the brain.

5.3 ECONOMY OF DESIGN OPTION

The behavioural prediction is that smaller animals will not have reduced behavioural capacities. The morphological prediction is that smaller animals will not have "special" morphological compensations that can be seen at the level of overall CNS volumes or the structures housing them. The compensations will occur instead at levels finer than overall brain size.

These options are not mutually exclusive. The basic questions are not theoretical, but rather empirical: Do very small animals tend to sacrifice their behaviour capacities due to allometric constraints on the relative sizes of their nervous systems? Do they make especially substantial sacrifices to devote unusual amounts of energy and material to build and maintain disproportionately large nervous systems? Or have they evolved special designs that allow them to reduce the amount of neural tissue needed to maintain particular behavioural capabilities? Testing these predictions against data will not determine which option is "correct" in a hypothetico-deductive sense, but will instead point to which solutions have been adopted by which lineages of animals that have faced the problems of miniaturization.

6 Data testing the predictions

Evolutionary adjustments in behaviour and morphology to miniature size have evolved independently in different lineages of insects and spiders, so we discuss data from the two groups separately.

6.1 MORPHOLOGY

6.1.1 *Insects*

Recent data on insects and spiders (Beutel *et al.*, 2005; Douglass *et al.*, in preparation; Grebennikov, 2008; L. Jiménez, unpublished; Polilov, 2005; Polivov and Beutel, 2009; Quesada *et al.*, submitted; Seid *et al.*, 2011; Wehner *et al.*, 2007) indicate that Haller's Rule also holds for arthropods, though they are typically much smaller in size than vertebrates (Fig. 1A). Thus, the brains of smaller insect species are smaller in absolute terms, but are proportionally larger than those of larger insects (Fig. 1B). In addition to being proportionally large, the brains of smaller insects may have smaller neurons that are more densely packed (Beutel *et al.*, 2005; Wehner *et al.*, 2007). The tiny first instar larva of the strepsipteran *Mengenilla chobauti* (body about 0.2 mm long), has a brain (supra- and sub-esophageal ganglia) that, relative to its body volume, is ~250 times larger than that of a large water beetle (Beutel *et al.*, 2005). The brain of the first instar larva of a tiny beetle, *Mikado* sp. (body mass, approximately 0.0016 mg—see caption of Fig. 1), constitutes about 16% of its body mass (Polivov and Beutel, 2009), while that of a tiny ant, *Brachymyrmex* sp. (body mass, ~0.04 mg), constitutes approximately 15% of its biomass (Seid *et al.*, 2011); for comparison, the brain of a large beetle constitutes only about 0.02% of its body mass, and a human brain constitutes only about 2% (Beutel *et al.*, 2005).

One set of dramatic morphological features of very small insects that fit predictions of the over-sized brain option concern the morphological distortions that are apparently associated with fitting their brains into limited space in their head capsules. For instance, in both tiny first instar strepipteran larvae and tiny ptiliid beetle larvae, the brain extends into the prothorax (Fig. 4). In the strepsipteran, *Mengenilla* sp. "[t]he whole central nervous system is extremely compact and completely shifted to the posterior thoracic region and abdomen (Beutel *et al.*, 2005, p. 308)." The brains of these tiny animals are so disproportionally large that they no longer fit where they are normally housed (the head capsule), and extend into other nearby portions of the body.

6.1.2 *Spiders*

In *Argiope aurantia* the relative brain size of tiny spiderlings is more than 10 times that of adults (compared with cephalothorax volume), and the proportion of a spiderling's brain mass dedicated to cortex versus neuropil was double that for the adult (Babu, 1975) [Here, we use "brain" to refer to the combination of supra- and sub-esophageal ganglia in spiders (see Table 3); Babu uses "brain" to refer only to the supra-esophageal ganglion]. Further data on relative brain size for 10 species in four additional orbicularian families, which included spiders substantially smaller (<0.005 mg body mass) than those studied by Babu (1975), showed a pattern of brain–body scaling consistent with Haller's

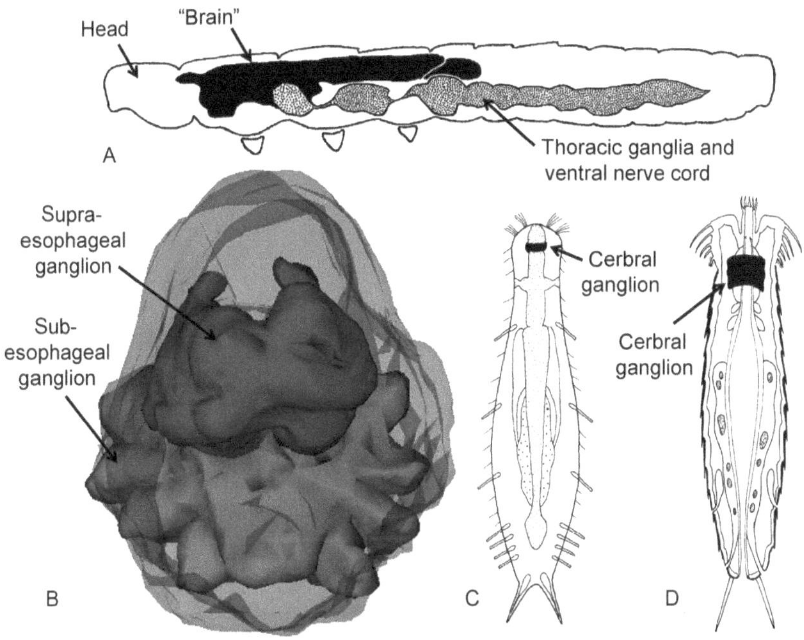

FIG. 4 Morphological correlates of grade changes. The larval beetle (A) and the spider nymph (B) are among the smallest representatives of their respective groups (they are 0.5 and 0.6 mm long respectively, and weigh about 0.002 and < 0.005 mg respectively). They both show signs of running out of room in which to house their relatively gigantic nervous systems. The beetle's brain is larger than its entire head, and has moved into its thorax and abdomen; in the spider, approximately 78% of the cephalothorax is occupied by the brain, which extends into the coxae of the legs and also bulges ventrally (not visible). In contrast, kinorhynchs (C) and gastrotrichs (D) are similar or even smaller in size (respectively they are about 0.3 and 0.2 mm long, and weigh approximately 0.002 mg), but their brains are relatively modest in size. (A from Polivov and Beutel, 2009; B from data in Quesada *et al.*, in review; C and D after Barnes *et al.*, 1993).

Rule (Fig. 1; Quesada *et al.*, in review). As with insects, smaller spiders also have smaller diameter neuron cell bodies (*op. cit.*). There was no obvious difference between the scaling of brain size of three kleptoparasitic species in this group and that of seven orb-weaving species.

The brains of very small spiders showed morphological distortions analogous to those in tiny insects. In this case, the brain extended into the coxae (Fig. 5), a design never seen in the larger spiders of this same study (Quesada *et al.*, in review) or in other species (Babu and Barth, 1984; Comstock, 1967; Hill, 1975; Kaestner, 1968). In addition, the relatively large brains of tiny spiders were associated with an apparent deformation of the sternum to accommodate the brain. The sternum of a first instar nymph of *Leucauge mariana* has a conspicuous ventral bulge, in contrast to the nearly flat sternum of the mature female

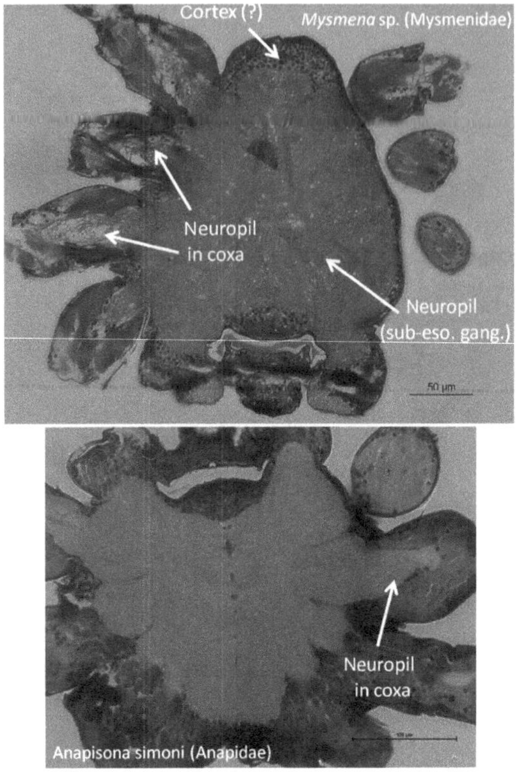

FIG. 5 Sections of tiny spiders showing extensions of the brain neuropil into their coxae. (See Color Plate 3 at the back of the book.)

(Fig. 6; Quesada *et al.*, in review). In serial sections, nearly the only tissue that occupies this bulge was the brain (the nerve cell bodies of the sub-esophageal ganglion). Tiny adult spiders in a variety of families, and tiny nymphs of species with medium and large-sized adults, also have ventrally bulging sterna (R. Madrigal, in preparation). Internal deformation is suggested in the early instar nymphs of the salticid *Phidippus clarus*; the digestive cecae that are abundant in the cephalothorax of the adult are nearly completely missing from the cephalothorax of a second instar nymph, which is instead largely occupied by the brain (Hill, 1975).

6.1.3 *Other invertebrates*

The huge relative sizes of the brains of tiny insects and spiders, and the deformed shapes of their brains and the body structures that house them (Fig. 4A and B), lead to speculation that relative brain size may impose lower limits on the body sizes of these groups (Beutel *et al.*, 2005; Grebennikov,

FIG. 6 Lateral views of an early instar nymph (left) and an adult female (right) of *Leucauge mariana* (legs removed), showing the distinctly greater ventral bulge of the sternum of the nymph.

2008). Extrapolating from arthropod brain allometries (Fig. 1), an animal that is an order of magnitude smaller than minute arthropods (e.g. on the order of 0.001 mg in weight) would have a brain that is about 25% of its body weight, a proportion which is seemingly prohibitive (Fig. 1B). Nevertheless, there are many animals whose adults are substantially smaller than the smallest insects and spiders, including tardigrades, gastrotrichs, kinorhynchs, rotifers, nematodes and some crustaceans (''small invertebrates'' in what follows), and the immature stages of these groups and others are smaller still.

How do the relative brain sizes of small invertebrates compare with those of insects and spiders? Measurements of the weights of their brains and bodies that would allow direct comparisons with data from other groups are generally not available. We have instead used published drawings of the nervous systems of a few small invertebrates with geometrically regular forms to obtain preliminary answers to this question. The answer we have found, that the brains of at least some are remarkably *un*exceptional in relative size and design (Fig. 4C and D), may eventually help illuminate the significance of the data from other taxa.

As a first approximation, we used classic two-dimensional drawings illustrating dissections, and deconstructed the bodies and brains of a few morphologically simple animals into multiple geometrically simple objects whose volumes could be easily calculated. For example, many kinorhynchs are more or less rectangular in dorsal view (with the introvert withdrawn; Fig. 4D); the body is ''... flattened ventrally, arched dorsally,'' and ''... triangular or nearly so in cross-section'' (Higgins, 1983; it forms an isoceles triangle with corner angles of 47° in *Kinorhynchus stenopygus* in Fig. 296 of Higgins, 1983). The area enclosed in a tracing of this triangular outline from the figure was 42% that of a circle whose diameter was equal to the length of the base of the triangle (i.e. the width of the animal in dorsal view). The brain has the form of a thickened sheet that surrounds the muscular pharynx, which is also triangular in cross-section (Barnes *et al.*, 1993). Similar volume estimates were made by deconstructing a nematode, a mite, and the first nauplius larva of a small copepod (see legend of Fig. 1).

Relative brain sizes in the tiny invertebrates are substantially smaller than would be predicted by simple extensions of the allometric lines of insects and spiders (Fig. 1). In terms of overall morphology, their nervous systems occupy only moderate fractions of their body volumes (Fig. 1B), and they show no morphological signs of being physically constrained by the volumes of the areas of their bodies where they are housed (Fig. 4). These species are typical rather than exceptional among very small invertebrates in not showing signs of disproportionately large brains, as judged by published drawings of gastrotrichs, tartigrades, ostracods, loriciferans, copepods and nematodes (Fig. 4; Barnes *et al.*, 1993; Brown, 1950; Brusca and Brusca, 1990; Kristensen and Higgins, 1984; Stachowitsch, 1992).

It is not clear whether the displacement of allometric lines to the left (Fig. 1B) in these groups is associated with behavioural inferiority and conforms to the size limitation option, or whether their CNSs possess economies of design that permit greater behavioural capability for a brain of a given size. Given the huge numbers of neurons thought to occur in even the tiny insects (e.g. Beutel *et al.*, 2005), it is striking that some of these other, smaller invertebrates function with very small numbers of neurons—302 in the hermaphroditic form of the nematode *C. elegans* (Altun and Hall, 2008), approximately 254 in the nematode *Ascaris lumbricoides* (Bullock and Horridge, 1965), and perhaps 300–400 and certainly less than 1000 in the nauplius larva of the copepod *Dactylopusia* (T. Lacalli, personal communication).

6.2 BEHAVIOUR

6.2.1 *Insects*

What is known about the behavioural capacities of tiny insects? Cole (1985) argued persuasively that several previous studies that attempted to relate behavioural complexity and brain size (Eisenberg, 1981; Eisenberg and Wilson, 1978; Harvey *et al.*, 1980; Howse, 1974) suffered from a lack of objective criteria for quantifying behavioural "complexity." Although it may be intuitively obvious that some animals have more complex behaviour than others (e.g. a limpet vs. a monkey), we lack methods to objectively characterize behavioural complexity for different taxa, or even the different developmental stages of a single species, in ways that are biologically meaningful (see Tononi *et al.*, 1994 for methods to describe brain complexity). This problem of crafting an operational definition of behavioural complexity resembles that faced by U.S. Supreme Court Justice Potter Stewart when trying to develop objective criteria to identify "hard-core pornography"; Stewart (1964) admitted that he could perhaps never objectively establish such criteria, but claimed that "I know it [pornography] when I see it.'' Indeed, the concept of ''behavioural complexity'' is so intuitively appealing it continues to be invoked, even though it is vague and unquantifiable in a biologically useful way. We have no confidence we can objectively distinguish pornography, and likewise believe that ''behavioural complexity'' is best abandoned due to lack of objective criteria.

''Behavioural complexity'' was the focus of studies of ants that concerned inter-specific or inter-caste differences to assess the behavioural consequences of miniaturization, but these studies yielded unconvincing and inconsistent results (Cole, 1985; Wilson, 1984). In the first place, they relied on a questionable method of quantifying behavioural complexity—the size of the behavioural repertoire (number of tasks) of a species. Cole (1985) found a positive correlation between head size and behavioural repertoire in nine species of as many genera, and concluded that the behaviour of smaller species of ants is less complex. This conclusion suffered from problems in quantifying ''complexity,'' including the uncritical acceptance of several untested assumptions: all tasks are equally demanding with respect to neural capabilities; tasks that are assigned the same name in different species (e.g. ''feed the larvae'') are equally demanding; the rate of errors in the performance of a given task does not vary between species; head size correlates with brain size in the same way in different ant species; all species respond to laboratory environments in the same way *viz-à-viz* expression of behaviour; and different investigators categorized behaviour in the same way, with no differences in lumping or splitting among different observers. In addition, appropriate controls for phylogenetic inertia (Harvey and Pagel, 1991) were lacking. A number of more recent studies have followed Cole (1985) and used size of behavioural repertoire as a metric of behavioural complexity, but suffer from the same problems (e.g. Anderson and McShea, 2001; Changizi, 2003; Chittka and Niven, 2009).

Other studies appeared to show the opposite correlation between small size and behavioural complexity.'' The brains of termite species were said to show the opposite correlation—species with greater behavioural ''complexity'' (in terms of nest architecture) have smaller brains with fewer neurons (Howse, 1974). Comparative studies of 10 species of *Pheidole* ants showed no correlation between repertoire size and body size in minor workers, and a weak (non-significant) negative correlation in majors (Wilson, 1984). These studies suffered from the problems in quantifying complexity described above.

Some authors have argued against the size-limitation hypothesis in insects on the basis of only negligible behavioural data. Beutel *et al.* (2005) noted that tiny strepsipteran larvae have several types of sense organs, can jump, and are able to locate hosts; they speculated that these larvae may have no behavioural impairments compared with larger insects, but provided no further data. Polilov (2005) cited unspecified indications that the behaviour of very small mites (0.08 mm long) is simpler than that of larger ones. Polivov and Beutel (2009) wrote that the tiny first instar larva of *Mikado* sp. possesses ''. . . a complex behavior pattern . . .,'' but gave no supporting citations or data.

In general, detailed studies of the behaviour of tiny insects are lacking, and it is presently not possible to assess whether the behavioural capabilities of tiny animals are generally affected by brain-related scaling problems, such as increased error rates, slower execution, greater bias toward computationally less demanding movements such as jumping (as opposed, for example, to reaching for objects in different directions), or other possible manifestations of reduced performance. In sum, there are far too few data to evaluate the behavioural limitation hypothesis in insects.

6.2.2 *Spiders*

Orb-weaving spiders are especially attractive subjects for studies that address questions related to size in behavioural and neural system trade-offs. In the first place, they vary widely in size, spanning more than five orders of magnitude in weight, from < 0.005 mg in newly emerged mysmenid and anapid nymphs to > 2000 mg in mature female nephilids (Quesada *et al.*, in review). Secondly, fine details of their web-building behaviour are especially easy to study, because orb construction produces a detailed and exquisitely precise record of hundreds of behavioural decisions and the contexts in which they occurred (Witt *et al.*, 1968). Even the webs of tiny individuals can be easily recorded and analysed in photographs (Fig. 7; Eberhard, 2007). Many features of an orb are readily quantifiable, allowing unusually detailed analyses of behavioural decisions. Some behaviours are repeated over and over during the construction of each web, thus allowing analyses of both the responses to particular stimuli, and of the precision or repeatability of these responses. This helps solve the difficult

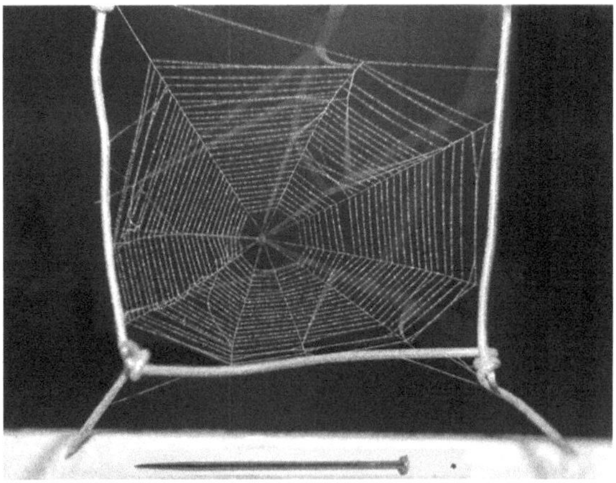

FIG. 7 Montage of photos (all to the same scale) of the web of an early instar *Anapisona simoni* built in captivity, a sewing pin, and the spider that built the web (the dark spot to the right of the pin).

problem of devising meaningful metrics to compare behavioural capabilities among diverse groups of animals (Healy and Rowe, 2007; Tinbergen, 1951). The other side of this coin is, of course, that only certain types of behaviour are included: web construction represents only a sample of a spider's behavioural capabilities.

In addition, details of many behaviour patterns used to build orbs are phylogenetically quite conservative (Eberhard, 1982; Scharff and Coddington, 1997), and are very similar in species with widely different body sizes. Ontogenetic changes in external morphology and predatory behaviour are typically small in spiders, so especially powerful intra-specific comparisons can be made that span large size ranges, thus obviating possible biases associated with phylogenetic inertia that can affect inter-specific comparisons (Felsenstein, 1985; Harvey and Pagel, 1991). The higher level phylogeny of spiders, especially in the group Orbiculariae, is relatively well studied (Griswold *et al.*, 1998; Kuntner *et al.*, 2008; Scharff and Coddington, 1997), and it is possible to identify cases in which miniature body size is secondarily derived in groups that have comparable web designs.

These advantages of orb weavers have begun to be exploited (Eberhard, 2007, 2011; Hesselberg, 2010), providing the most extensive and detailed comparisons of the behaviour of tiny and moderately sized arthropods currently available. These results are discussed below, and summarized in Table 1.

6.2.2.1. Behavioural precision Due to the problems in quantifying ''behavioural complexity'' (see Section 6.2.1), we focus mostly on two other variables that could reflect possible behavioural limitations: behavioural precision—the animal's ability to repeat the same behaviour precisely; and flexibility—the animal's ability to adjust its behaviour appropriately to different environmental circumstances. Both traits have been hypothesized to be less developed in animals with smaller brains (Misunami *et al.*, 2004), because they would presumably require sensors and processors to perform consistent behaviour in different circumstances, to assess environmental conditions, and to appropriately trigger and adjust expression of alternative behaviours.

The logic of using behavioural precision is the following. There are several reasons to suspect that motor behaviour, coordination, and orientation may be less precise in an animal with a reduced nervous system, particularly when the numbers of neurons are reduced (Grebennikov, 2008). Smaller animals have fewer sense organs, and thus have less complete sensory input (see Section 1). They may also perform less thorough or precise analyses of sensory inputs, due to lower numbers of inter-neurons, fewer dendrites, or fewer or more noisy connections between them (Chittka and Niven, 2009; Faisal *et al.*, 2005; Wittenberg and Wang, 2007). If they have less extensive internal feedback mechanisms, the behavioural imprecision that results from intrinsic noise in the nervous system may also be greater (Calvin and Stephens, 1967, 1968, Eberhard, 1990a, 2000). The smaller numbers of motor axons and reduced

TABLE 1

Tests of the size-limitation hypothesis using the behaviour and morphology of orb-weaving spiders, utilizing behavioural data from (in order of increasing size) second instar nymphs and adults of *A. simoni*, second instar nymphs and adults of *Allocyclosa bifurca*, and adults of *Leucauge mariana* (from Eberhard, 2007, 2011), and morphological data from these and other species (from Quesada *et al.*, submitted; Vargas, unpublished)

Predictions of the size-limitation hypothesis for smaller spiders	Finding	Support hypothesis? (Comments)
Behaviour		
Reduced adaptive flexibility		
A. Lack of correlation between number of radii and number of sticky spiral loops	Correlations occur in spiders of all sizes	No
B. Lack of correlation between angles between adjacent radii and their lengths	Correlations occur in spiders of all sizes	No
C. Lack compensatory reduction in sticky spiral spacing following an over-sized space	Lack compensatory spacing lack in smallest species; no lack in small nymphs of another species[a]	Yes, but Weber-Fechner Law offers alternative explanation for lack of compensatory space in the smallest species
Reduced behavioural precision		
D. Greater imprecision in spacing of sticky spiral	No correlation between spider size and imprecision	No
E. Weaker correlation between number of radii and number of sticky spiral loops	Weaker correlation in smaller species but no difference between conspecific nymphs and adults[b]	Mixed[b]
F. Weaker correlation of the lengths of adjacent radii and the angle between them	No intra-specific differences between large and small spiders[c]	No[c]
G. Larger differences between adjacent spaces of sticky spiral	No trend with body size between or within species	No
H. Weaker correlation between numbers of upper radii and sticky lines above the orb	No difference between nymphs and adults of *A. simoni*	No

(*continues*)

TABLE 1 (*Continued*)

Predictions of the size-limitation hypothesis for smaller spiders	Finding	Support hypothesis? (Comments)
Reduced "complexity"		
I. Simpler design of orb	Smallest species has lost temporary spiral	Yes[d] (but the selective advantage of losing the temporary spiral offers an alternative explanation[d])
J. Single web design	Only smallest species has alternative web design	No (opposite)
K. No modification of design according to available space	Only smallest species modifies orb design (angle of cone) to available space	No[e] (opposite) (larger species have only two-dimensional webs, so such adjustments are not feasible)
L. No correlation between radial and sticky lines above the orb	Only smallest species modifies orb on basis of number of lines above the orb	No[e] (opposite) (larger species do not have sticky lines above the orb)
Morphology		
M. Lack of distortion of body design to house "overly large" brain	tiny spiders have bulging sterna and sub-esophageal ganglion extending into coxae; tiny insects have brains extending into prothorax; tiny mites have brain extending into coxae	No

[a] Also occur in webs of the moderately large araneid *M. duodecimspinosa*

[b] Scatter around the regression line was significantly greater in the tiny species *A. simoni* than in larger species, and in small nymphs than in adults of *A. bifurca*. This support for the size-limitation hypothesis was inconsistent, however, because the scatter in nymphs of *A. simoni* was not greater than that in conspecific adults.

[c] Because the behaviour thought to be responsible for producing this correlation occurs during the production of "secondary" but not that of "primary" radii (Eberhard, 1990b; Hingston, 1920; LeGuelte, 1967), and because the webs of different species in this study have different mixes of primary and secondary radii (below), it is appropriate to test for size effects by making intra-specific rather than inter-specific comparisons. Comparisons in both *A. bifurca* and *A. simoni* showed nearly identical amounts of scatter around regression lines for nymphs and adults (Eberhard, 2011).

[d] Loss of the temporary spiral may be an adaptation to allow construction of sticky lines above the plane of the orb as part of sticky spiral construction (Eberhard, 2011).

[e] Direct comparison between present-day larger and smaller spiders is not possible, but evolutionary prediction of reduction in behavioural abilities when evolve small size is contradicted.

feedback could also result in less subtle control of body movements. In an animal like a web-building spider, which uses the movements and positions of its own legs to provide sensory input (Barth, 2002), limitations in the precision of leg movements and in the proprioreceptive analysis of these movements could significantly reduce the accuracy of its sensory perception of the world.

Construction of the sticky spiral in an orb is well-suited to analyses of behavioural precision. The spider starts near the periphery of the orb and works inward, repeating some processes such as those illustrated in Fig. 8 at each radius it crosses; the space between loops of sticky spiral is determined each time the spiral is attached to a radius. During the construction of an orb the spider repeatedly (often hundreds of times) senses the location of the inner loop of sticky spiral, measures the distance along the radius at which the current loop should be attached, and then attaches it.

Simulation studies have shown that a relatively simple set of instructions can produce an at least superficially spider-like sticky spiral pattern (Eberhard, 1969; Krink and Vollrath, 1999). Nevertheless, control of the behaviour that determines the spacing between sticky spiral loops is affected by at least six different factors: the length of the exploratory leg I (Reed *et al.*, 1965; Vollrath, 1987); the angle of the radius with gravity (LeGuelte, 1967; Vollrath, 1992; Witt *et al.*, 1968); the distance of the attachment from the hub (LeGuelte, 1967; Vollrath, 1992); the age of the spider (Reed *et al.*, 1969); an internally determined ''set point'' that is influenced by silk supplies (Eberhard, 1988a); and the distance of the previous sticky spiral attachment from the outer loop of temporary spiral (Eberhard, 2011; Eberhard and Hesselberg, in review).

Sticky spiral construction behaviour is highly conserved taxonomically (Eberhard, 1982; Griswold *et al.*, 1998). Sticky spiral placement behaviour involves only relatively simple behavioural actions, and perhaps also relatively low level analyses of stimuli by the spider, so it can only test a relatively strong version of the size-limitation hypothesis: that even relatively undemanding behaviour show limitations in tiny animals (more demanding behaviour might be more likely to be limited by size—see below).

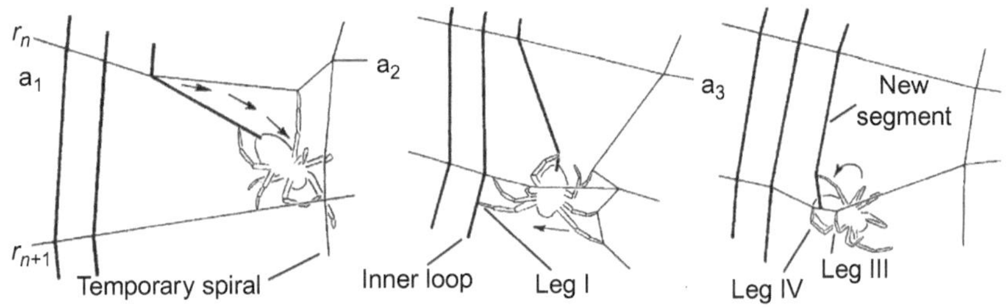

FIG. 8 Processes that are repeated up to hundreds of times during the placement of the sticky spiral in an orb web.

The expectation of the size-limitation hypothesis is that very small spiders will show greater imprecision in sticky spiral construction behaviour. This would result in greater differences in the adjacent spaces between loops of sticky spiral. The highly regular spaces between successive loops of sticky spiral in orb webs imply that, at least in a given area of the orb, a particular spacing is advantageous (Eberhard, 1986; Witt, 1965); thus variation between neighbouring spaces probably result from mistakes. The validity of using behavioural comparisons between young and old spiders of the same species to test for possible handicaps of miniaturization depends on the assumption that learning does not reduce variation. This assumption is supported by the repeated finding that juvenile experience does not influence major aspects of orb construction (Eberhard, 2007; Mayer, 1952; Reed *et al.*, 1970; Vollrath, 1992). An additional possible consequence of size limitation might be that the speed of execution of given behaviour patterns would be reduced in smaller individuals. The expectation of the size-limitation hypothesis is thus that to achieve a similar level of precision, tiny spiders might work more slowly.

Comparing first instar nymphs and adults from three different spider species that spanned a range from 0.6 mm body length and < 0.005 mg wet weight (first instar nymphs of the anapid *Anapisona simoni*) to body length about 7 mm and wet weight about 50–80 mg (adult females of the tetragnathid *L. mariana*), the smallest individuals showed no more imprecision than did the larger spiders (Eberhard, 2007; Table 1) (the precision of sticky spiral spacing was defined as a measure of intra-individual imprecision in the spacing of successive loops of sticky spiral). The imprecision values of the tiniest spiders were in fact significantly smaller than those for the larger second instar nymphs of *Allocyclosa bifurca*, and were not significantly different from those of the much larger adults of *A. bifurca* and *L. mariana* (Eberhard, 2007).

The importance of these data as evidence against the size-limitation hypothesis is emphasized by two further considerations. Anapids must travel much larger distances from one attachment of sticky spiral to the next during sticky spiral construction (in relation to their body size) than do most orb weavers (Fig. 9). The radial distances from the hub inward and back out are used as cues to direct sticky spiral spacing (Eberhard and Hesselberg, in review; Krink and Vollrath, 1999), and temporary spiral placement (Eberhard, 1988b). Although the psychophysics of spider perception is not well understood, general psychophysical principles should hold for them as well. The Weber-Fechner Law (in the sense used by Cheng *et al.*, 1999) holds that an animal's ability to detect a just noticeable difference (JND) for a given stimulus is a function of the magnitude of the stimulus, and the JND is a constant proportion across a large range of magnitudes (Shettleworth, 2010; Stevens, 1975). A consequence of this law is that small things are assessed more precisely in absolute terms than larger things. Thus, a measurement made by a spider walking a longer distance along a line is

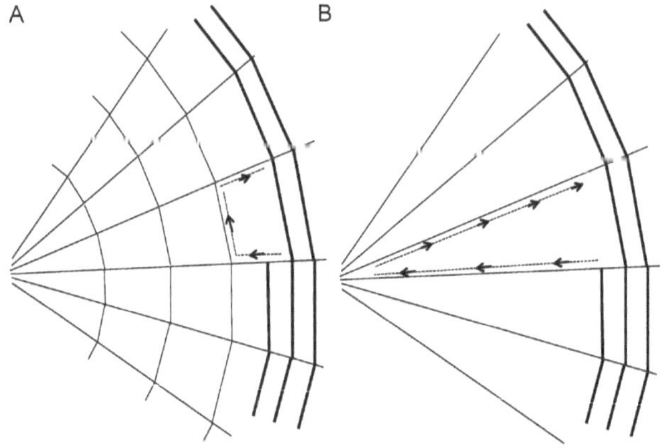

FIG. 9 Paths taken by the spider in moving from one attachment of the sticky spiral to the next during sticky spiral construction by a typical orb weaver (left) and by an anapid, whose web lacks a temporary spiral (right).

expected to be less precise than one measured over a shorter distance. This implies that *A. simoni* is less likely to be able to sense small differences between the relatively long distances it must travel inward and outward during sticky spiral construction (Eberhard, 2011). In short, despite the probable greater difficulty for the anapids to sense important cues, they showed comparable or better precision in sticky spiral spacing.

A second extenuating fact is that the webs of anapids have fewer loops of sticky spiral. In many species the sticky spiral spacing is different at different distances from the hub. In comparing the space between two loops of sticky spiral with the spaces between adjacent loops of sticky spiral, this distance from the hub is more likely to introduce variation in anapid webs and cause their precision values to be higher.

Another size-limitation prediction—that smaller spiders move relatively more slowly—was also not met. The mean times between successive attachments of the sticky spiral during the early phase of web weaving were 3.0 s in a small individual of the smallest species, *A. simoni*; 1.9 and 3.4 s in a young nymph (1.4 mg) and an adult female (30 mg) of *A. bifurca*, respectively, and 1.5 and 7.3 s in mature females of *L. mariana* (50 mg) and the araneid *Gasteracantha cancriformis* (100 gm), respectively. The smallest spiders had to move much farther between attachments with respect to their body size (Fig. 9, Eberhard, 2011), and they were thus moving much *more* rapidly in terms of their body size.

Imprecision may arise from three different sources: imprecision in what the spider *intended* to do (e.g. due to conflicting motivations or variable conditions); imprecision in the control of the behavioural movements that were involved in executing these plans; and imprecision in executing other, previous behaviour patterns that, as a secondary consequence, influence execution of the behaviour in question. Data are generally lacking to assess the relative importance of these possibilities. A possible example of the first type of effect (variation in intention) would be the association recently found in sticky spiral spacing by *Micrathena duodecimspinosa* between over-sized sticky spiral spaces and the distance to the nearest temporary spiral scar (Eberhard and Hesselberg, in review). In this species, imprecision in sticky spiral spacing may be associated with inconsistency in the influence of occasional sharp changes in the distance from the outer loop of temporary spiral that "distract" the spider, causing it to pay less attention to other cues. This species also provides a possible example of the third source whereby imprecision arises as a consequence of imprecision in other behaviour. Sticky spiral spacing also seems to be influenced by the presence of intact temporary spiral lines which the spider failed to break during sticky spiral construction (W. Eberhard, unpublished data). In this case, the apparent cause of some of the imprecision in sticky spiral spacing is the occasional previous failure to cut the temporary spiral during sticky spiral construction.

6.2.2.2. Behavioural flexibility and the precision of adjustments among web variables It may be that the precision of small spiders just discussed involves, at least on the motor side, behaviours that are relatively undemanding in terms of nervous system capacities, and that it thus constitutes an overly-stringent test of the size-limitation hypothesis. Placement of a new segment of sticky spiral will depend on where legs III and IV grasp the radius relative to the site where exploratory taps with leg I contacted the inner loop, as the site where the spinnerets attach the sticky spiral line to the radius is between these legs (Fig. 8). Consistency in sticky spiral spacing is thus likely to be, at least to some extent, a direct result of the size of the spider's own body (Vollrath, 1987). Perhaps the behavioural limitations of tiny spiders are only manifest in the execution of more neurobiologically challenging tasks. Study of more challenging behaviour patterns might thus give more sensitive tests of the size-limitation hypothesis. Such tests could involve the abilities of spiders to adjust one web variable to another, or of the precision with which such adjustments are made. The results of several such tests are summarized in Table 1, and discussed below.

Some tests involved the degree to which spiders of different sizes flexibly adjusted one web variable to another, as indicated by the existence of a significant correlation between the two variables, and the precision of these adjustments. Precision was measured by the tightness of this relationship, in terms of the residuals around a regression of one variable on the other. The general patterns in both the existence and the precision of these adjustments failed to fit size-limitation predictions (Table 1). Several of the correlations are probably selectively advantageous (summary in Eberhard, 2011).

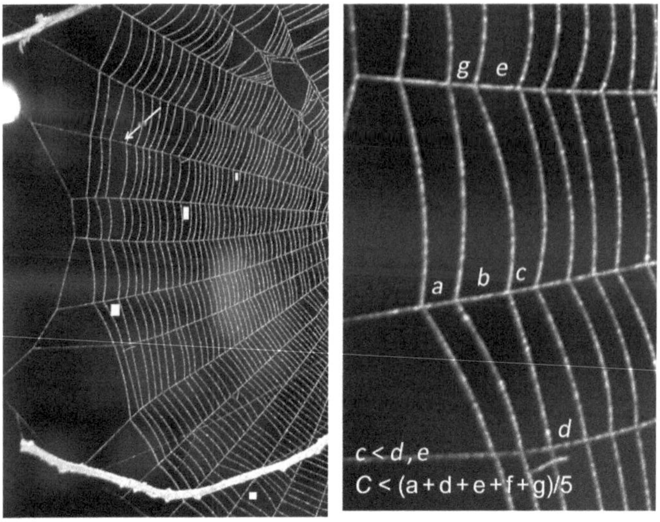

FIG. 10 An over-sized space between loops of sticky spiral (*b*) that was followed by a reduced space (*c*) in the web of an adult female *Micrathena duodecimspinosa*. The two inequalities indicate ways to calculate whether or not a spider reduced the space (*c*) that followed an over-sized space.

The trend to produce compensatory reductions in sticky spiral spacing following over-sized spaces in orb webs (Fig. 10; Table 1C) has only recently been recognized (Eberhard, 2011), but it may also be a general pattern in orb webs (Eberhard, unpublished data). Compensatory adjustments may function to reduce irregularities in the path of the inner loop of sticky spiral during sticky spiral construction, and help re-establish the uniformity of the spacing of sticky spiral loops after a discontinuity. They are probably triggered when the spider senses differences in the distances travelled inward and outward along radii as it builds the sticky spiral (Eberhard and Hesselberg, in review). This adjustment was lacking in the smallest species (Table 1).

6.2.2.3. Qualitative differences in behavioural flexibility: adding and subtracting abilities Although quantitative metrics for behavioural complexity are problematic (see Section 1), there are several qualitative differences in which the behavioural capacities of different-sized spiders can be compared.

The number of upper radii in the webs of the smallest species, *A. simoni* (Fig. 11), showed a positive correlation with the number of sticky lines attached to any single upper radius in the webs of both adults and small nymphs (Table 1 L). None of the other species built sticky lines out of the plane of the orb, and the plesiomorphic state for orb weavers is planar orbs (Griswold *et al.*, 1998), so the ability of *A. simoni* to adjust the sticky lines to upper radii is

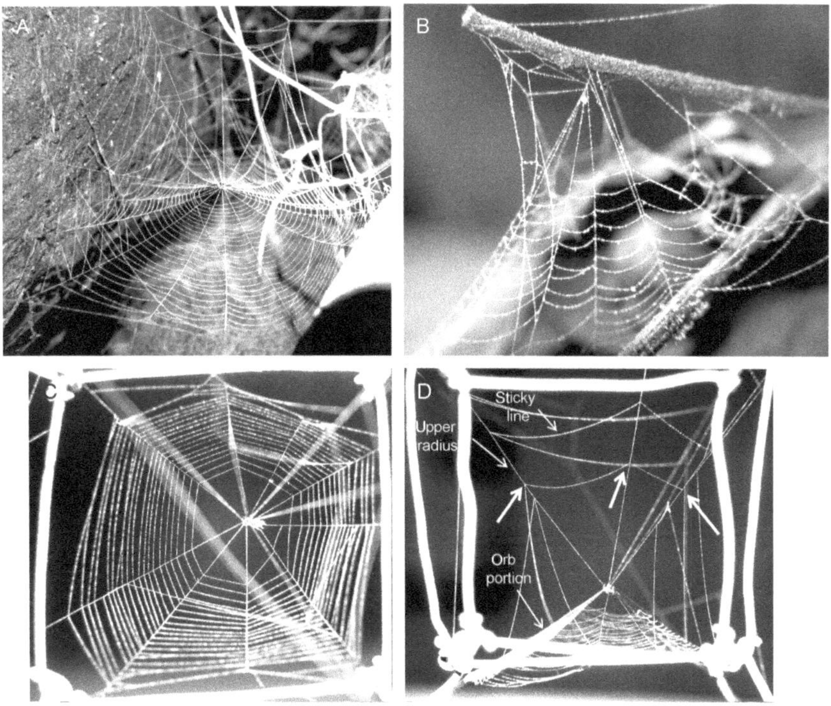

FIG. 11 Lateral views of webs of *Anapisona simony* (A, D), showing radii above the plane of the orb and sticky spiral lines attached to these radii (dorsal view in C). The alternate web design in *A. simony* is shown in B. The radii of an incomplete orb converge in an area near the underside of a twig where the spider rests, but do not form a hub.

derived. Similarly, when adult *A. simoni* were offered two different sizes of wire cubes in which to build their webs, the hub of the orb portion of the web was drawn upward more sharply when the spider built in a smaller cube (Table 1K). This modification of the orbs in smaller cubes increased the area covered by the catching spiral over the area it would have had if the orb had been more nearly planar. No such adjustments occurred in the webs of the other, larger species, which built only planar or nearly planar orbs.

Still another additional ability in *A. simoni* was for the same spider to build either a normal orb or an alternative web form lacking a discrete hub and with the radii converging on the underside of a twig or other large object under which the spider rested (Fig. 11B; Table 1J). Construction of this alternative, ''sector web'' design probably involves alterations in frame, radius and hub construction behaviour (Eberhard, 2011). A similar sector web design occurs, along with complete orbs, in another, unidentified tiny anapid (Madrigal and WGE, unpublished data). Alternate forms of prey capture webs were never seen in samples of hundreds of orbs in the field in species with which we have extensive experience (the tetragnathids *L. mariana*, *Leucauge argyra*, and *Dolichognatha* sp.; the araneids *A. bifurca*, *M. duodecimspinosa*, *Micrathena sexspinosa*, *Metazygia gregalis* and *Araneus expletus*; and the uloborid *Uloborus diversus*), and are not mentioned in general reviews of orb webs (Vollrath, 1992; Witt *et al.*, 1968). We know of only four other cases, all of which are clearly independent (Eberhard, 1969, 1990c; Edmunds, 1978; Sandoval, 1994). Lack of alternative webs is plesiomorphic in orb weavers (Griswold *et al.*, 1998), so construction of alternative web forms in anapids is derived, suggesting these tiny spiders have additional behaviours relative to those of many larger spiders.

One final qualitative difference involves web features that are common in larger species yet lacking in minute ones: the orbs of the smallest species, *A. simoni*, lack temporary spirals, which are standard in the orbs of most other orb weavers (Table 1I). This support for the size-limitation hypothesis is weakened, however, by the fact that there is a likely alternative explanation. Omission of the temporary spiral permits the spider to extend its sticky spiral lines into the space above the orb, where they can aid in trapping prey passing above the plane of the orb (Eberhard, 2011).

6.2.2.4. General conclusions and summary of orb construction behaviour data In general, the behavioural data just reviewed contradict rather than support the size-limitation hypothesis that smaller animals are more limited in their behaviour (Table 1). They also contradict the more general-hypothesis of behavioural inferiority for species with brains of smaller absolute sizes (Byrne, 1997; Striedter, 2005). Both intra- and inter-specific comparisons lead to the same conclusions, so phylogenetic bias is probably not a problem in these analyses. To our knowledge, these are the most detailed comparative behavioural data ever assembled for animals with very small body sizes.

A simple tally of the numbers of contradictions and confirmations in Table 1 (nine against; one in favour; two with mixed support) is inappropriate, however, because these behavioural capabilities are not necessarily comparable. There are reasons to expect that some of these capabilities may be more costly or cheaper in terms of nervous tissue than others. It is possible to make educated guesses regarding what neuronal capacities are needed to confer the ability to adjust construction behaviour appropriately to different environmental conditions. Some adjustments may be relatively simple for a spider to accomplish. For instance, the radius angle could be adjusted to vary inversely with the length of the radius simply by always walking a fixed distance from the exit radius along the frame line before attaching the new radius (Hingston, 1920; though in fact this distance is not constant even in a given orb—Eberhard, 1972, unpublished data on *M. duodecimspinosa*). In contrast, it would seem much more difficult for a spider to assess the area available in which to build its orb, as it probably involves measurements of distances and directions travelled, and memories of both (Vollrath, 1992). This ability may be ancestral (Vollrath, 1992; Eberhard and Barrantes, in preparation). The ability of *A. simoni* to build more steeply sloping orbs to take better advantage of limited spaces (Table 1K) may also be ancestral, as it occurs in the larger, distantly related *Philoponella vicina* (Barrantes and Eberhard, unpublished data). The ability of *A. simoni* to correlate the number of upper radii and the number of sticky lines attached to them (Table 1L) also seems likely to be relatively demanding, and probably requires memory (Eberhard, 1987, 2011). The equal precision in sticky spiral spacing in *A. simoni* (Table 1D) weighs especially heavily against the size limitation option, because of the Weber-Fechner Law and the lack of temporary spirals in their orbs imply that this task is especially difficult for these spiders (Eberhard, 2011).

In addition, some of the evidence that supports the size limitation option is weak. The lack of compensatory adjustment of sticky spiral spacing following over-sized spaces in the anapid is likely a result of the loss of temporary spiral lines in the webs of this family.

As noted above, comparisons between conspecific nymphs and adults constitute a valid test of the effects of body size on imprecision only if learning has minimal or no effect on behavioural imprecision. This was indeed the case, as there were no significant differences between first instar nymphs and adult females in *A. simoni* or *A. bifurca*, and the insignificant trends that did occur were in opposite directions in both species. In sum, the evidence against the size limitation option is strong in the comparisons involving possibly more ''difficult'' tasks, strengthening the overall trend in the data against this hypothesis.

Rejection of the size limitation option would leave the ''over-sized brain'' and the ''economy of design'' options as possibilities for orb weavers. The behavioural data do not allow a choice between these options, which in any case are not mutually exclusive. The morphology of the brains of some of the smallest orb-weaving spiders, and that of the structures that house them (see Section 6.1.2), support the ''over-sized'' brain hypothesis. But until a broader survey of spiders is examined to determine the allometric line for the entire group, it is not known whether some of these spiders are relatively highly encephalized compared with others (see Section 7.1 below)

7 Discussion

7.1 MORPHOLOGY

''Haller's Rule'' is now well documented for both vertebrates and invertebrates (Fig. 1): within a grade smaller animals have relatively larger brains. Why this rule? We do not have an answer, but note that the wide taxonomic scope of the present discussion yields evidence against two of the hypotheses previously inspired by vertebrates. The idea that the rule results from scaling between body surface area and volume (Jerison, 1973) is not able to deal with the multiple between-grade differences, in which similar-sized animals of different taxa show sharply different nervous system allometries. Similarly, the hypothesis that metabolic constraints could explain the grade differences between poikilotherm reptiles and their homeotherm descendents, birds and mammals, does not explain differences between other poikilotherm groups such as ants, spiders and salamanders (Wehner *et al.*, 2007; Fig. 1).

A second way to visualize grade changes is to extrapolate the brain–body allometry slopes of different taxonomic groups. Figure 12 uses the summary graph of vertebrate brain–body allometries of Striedter (2005; Fig. 4.3B) as a base, and assumes that 20% of the body weight dedicated to the CNS is an absolute upper limit for a functional animal (the highest direct measures of these values that we know of are approximately 16–17% in ants and a beetle larva—see Section 6; also Fig. 1). Even with this relatively permissive assumption, the minimum feasible body size for a bird would be about 200 mg, for a typical mammal about 10 mg, and for a typical amphibian about 2 mg (Fig. 12). Obviously, many invertebrates are much smaller than these "limits." Similarly, nematodes and kinorhynchs have body sizes substantially smaller than would be feasible for animals with the body designs of insects and spiders. These grade changes involved solving scaling problems that were apparently insuperable for otherwise successful and hyperdiverse groups such as spiders and insects.

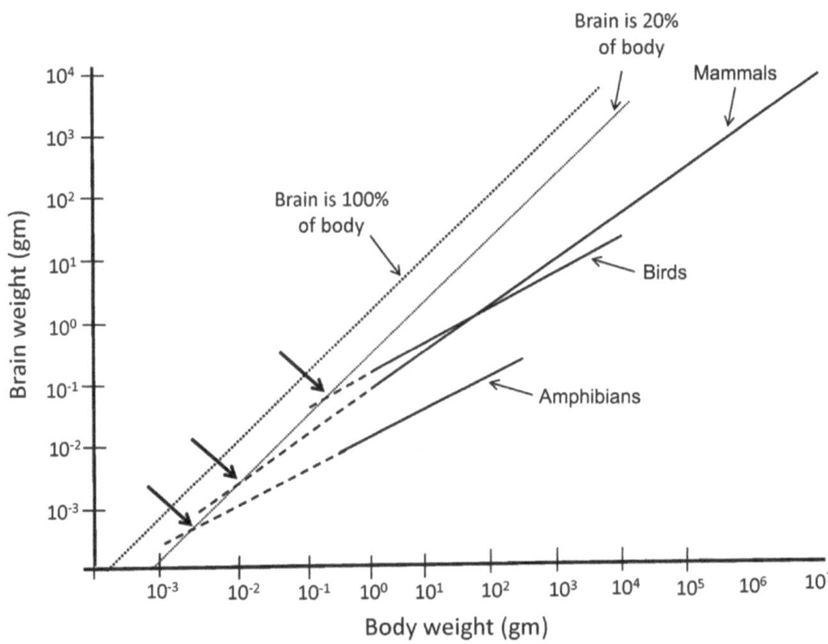

FIG. 12 Extrapolations (dotted lines) of allometric lines for mammals, birds and amphibians in Fig. 4.3B of Striedter (2005) (solid lines) show the expected inferior limits of body size in these groups (thick arrows) if one assumes that 20% is the maximum feasible amount of the body that can be dedicated to brain tissue.

Striking alterations of morphological design often occur in the smallest species in a taxon (grade). The brains of miniature salamanders are so large relative to their heads that some skull bones have been lost, apparently to make room for the brain (Hanken, 1983; Roth *et al.*, 1990). Similarly, the brains and structures housing them and their normal contents are deformed in very tiny insects, spiders and mites to accommodate the relatively large CNS (Beutel *et al.*, 2005; Grebennikov and Beutel, 2002; Quesada *et al.*, submitted; G. Vargas and R. Madrigal, unpublished; see Figs. 5, 6). The morphology of very tiny species of vertebrates, insects and spiders thus suggests that the smallest individuals in each group are near absolute group-specific lower limits in size. For instance, a study of the brains of miniature plethodeontid salamanders concluded that "visual system design may represent a primary limit to ... body size decrease," and that "further reduction of body size in these lineages, without an accompanying reduction in genome and cell size, may be possible only by impairment or even loss of visual function" (Roth *et al.*, 1990, pp 187, 188). Striedter (2005) and Grebennikov (2008) speculated that the vertebrate body plan, with the brain enclosed tightly in the cranial cavity, has imposed lower size limits due to limited abilities to accommodate larger brains by deforming the brain or the structures that house it. These limits do not seem to apply for other smaller taxa, in which even smaller animals have only "reasonably" sized brains (e.g. Fig. 4). Importantly, the external and internal deformations in tiny insects and spiders do not fit the prediction of unaltered form made by the size limitation and economy of design options.

At first glance it might appear that data supporting Haller's Rule (Fig. 1) imply that the smaller members of a given taxon have adopted the over-sized brain option. In fact, however, the log–log scales and the huge absolute ranges represented in Fig. 1 hide very substantial differences in brain–body allometries in different species within a grade. Many studies of vertebrates have correlated differences of this sort (e.g. the so-called encephalization quotients) with different behavioural capabilities (e.g. Striedter, 2005), and these differences may also be correlated with differences in economy of design. Similar variation in allometry between taxonomic groups is already apparent in insects (compare ants with weevils in Fig. 1), and it seems likely that future studies will find correlations with differences in their behavioural capabilities and economies of design within insects, spiders and other invertebrates.

It will be interesting to make further comparative surveys of scaling relationships within invertebrate taxa as has been done in vertebrates (Striedter, 2005). A study of 70 species of ants showed that there is a statistically significant break point in the allometric relationships relating brain volume to body mass near the lower end of the range of ant body size (Seid *et al.*, 2011). The slope for species below 0.9 mg body mass is steeper than that for larger species, and consequently the smallest species have brains that are *smaller* than expected if they followed the same slope as the larger species. This change in the statistical trend effectively dampens the strength of Haller's Rule, and may help reduce costs that would otherwise result from especially large brain sizes. Intraspecifically, a fungus-growing ant, *Atta colombica*, also showed a similar breakpoint, though at a higher body mass (1.4 mg); the smallest individuals again have brains that are smaller than would be expected if they followed the scaling relationship for larger sizes (Seid *et al.*, 2011). The behavioural significance of these break points is not known, and it is not clear whether the tiny ants with smaller than expected brains have reduced behavioural capabilities.

7.2 BEHAVIOUR

Efforts to understand the significance of Haller's Rule are impeded by the dearth of relevant behavioural data. The most detailed data come from only a single taxonomic group (orb-weaving spiders) and a single type of behaviour (web construction). It is clear, however, that the lack of size-imposed limitations in orb construction behaviour in the tiny *A. simoni* is probably not limited to this particular species. Other highly organized, three-dimensional derivatives of orbs also occur in species in other genera of Anapidae (Fig. 13), and also in the even smaller species in the related symphytognathoid families, Symphytognathidae and Mysmenidae (Coddington, 1986a; Eberhard, 1987; Lopardo *et al.*, 2010; Platnick and Shadab, 1979). Symphytognathoids include the smallest adult spiders of any species known, and their generally minute size is a derived trait with respect to other orb weavers (Griswold *et al.*, 1998). Orbs with sticky and non-sticky lines in the third dimension are widespread in symphytognathoids, where they apparently evolved once basally (Lopardo *et al.*, 2010). Some species even combine multiple orbs in the same web (Fig. 13). Superficial examination of photos of webs of other species in these families (Coddington, 1983, 1986a,b; Eberhard, 1987; Lopardo *et al.*, 2010; Platnick and Shadab, 1978, 1979) does not reveal obviously greater imprecision in the spacing of sticky spiral loops. In sum, the evolution of additional, complex web forms was associated with tiny body size, the opposite of expectations under the size-limitation hypothesis. Building a planar orb with additional lines that are oriented in consistent ways in the third dimension requires additional behavioural routines, which could be considered more "difficult." The association of small body size with three-dimensional additions to orbs that are behaviourally challenging is in the opposite direction from that predicted by the size-limitation hypothesis (Fig. 12).

Even very tiny orb-weaving spiders have not made detectable sacrifices in their behavioural capabilities, at least with respect to web construction, despite the likelihood that they are under intense selection to do so because of the relatively greater metabolic costs imposed by their relatively large brains (Quesada *et al.*, in review). Thus they pay the cost of these capacities by having proportionally gigantic brains (the "over-sized brain" option). Less complete data from other groups (ants, termites, vertebrates) do not show consistent

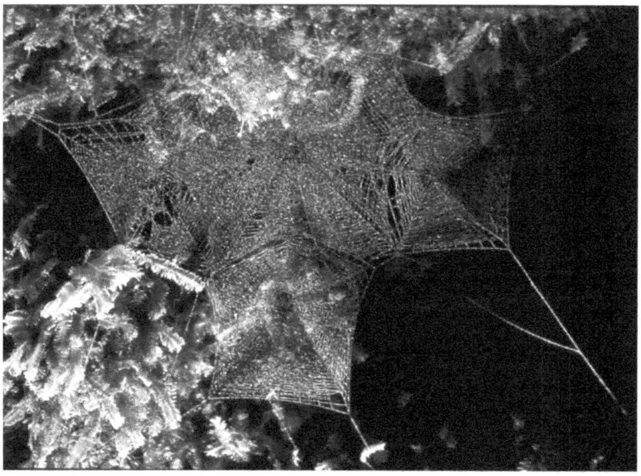

FIG. 13 The geometric complexity of the orbs of some tiny spiders is illustrated by this "quilt" web of a penultimate male of the tiny *Tasmanapis strahan* (approximately 1 mm long; like other tiny spiders, its sternum bulges ventrally, where part of its brain is housed); multiple orbs are apparently joined along their frame lines to form a single more or less horizontal planar structure, a composite design unknown in the many hundreds of species of larger spiders in the families Araneidae, Tetragnathidae and Uloboridae whose web forms are known (photo by L. Lopardo).

trends (Bonner, 2006; Coddington, 1986a,b; Howse, 1974; Rensch, 1960; Wilson, 1984).

It is not clear whether or not the evolutionary "refusal" of tiny orb-weaving spiders to sacrifice web construction behaviour represents a general trend in other groups of very small animals. It is possible that orb weavers are unusual in that they rely more heavily on behaviour to acquire resources than, say, a larval beetle that simply eats its way through the fungus where it lives. A preliminary, weak test did not support this idea: the brain scaling of three secondarily kleptoparasitic spider species was similar to that of their orb-weaving relatives (Quesada *et al.*, in review; of course, kleptoparasitism has its own behavioural requirements).

There are, however, very tiny insects that obviously also accomplish the same general tasks, such as finding food, avoiding predators and reproducing as those of larger insects. Some tiny insects make behavioural decisions that are apparently based on analyses involving multiple factors, such as the flexible searching and patch-leaving decisions of some parasitoid wasps that are influenced by both memory and internal physiology (reviews in Burger *et al.*, 2006; Hanson and Gauld, 2006; Outreman *et al.*, 2005; Vos *et al.*, 1998). It has not been tested, however, whether they perform their tasks less effectively (e.g. learn more slowly, with more mistakes, make fewer or less precise or subtle adjustments to environmental stimuli). The behavioural capacities of even tinier metazoans, such as mites, kinorhynchs, tardigrades, etc. are even less studied. Some polyphagous predatory mites are able to make relatively complex analyses of chemical cues from plants attacked by their prey, and to learn associations with different odours (summary in van Wijk *et al.*, 2006), thus providing a preliminary suggestion that the possible grade change in this secondarily miniaturized group (Fig. 1B) was not due to reductions in behavioural abilities.

The general failure of the predictions of the size limitation option for spiders does not mean that this option is not utilized by other groups. Nor does it mean that the logic of the arguments used to derive it is incorrect. The size limitation option is based on two suppositions, both of which are likely to be correct: (1) a smaller animal probably needs a relatively larger brain compared with its body size to maintain the same behavioural capabilities; and (2) a relatively large nervous system is especially expensive to build and maintain. The basic question is empirical, not theoretical: do very small animals tend to sacrifice their behavioural capacities due to allometric constraints on the relative sizes of their nervous systems? The answer from orb weavers seems to be ''No,'' but other groups will need to be studied to determine the generality of this answer.

7.3 THE MYSTERY OF ''GRADE CHANGES'' IN BRAIN ALLOMETRY

Figure 1 shows both that invertebrates follow Haller's Rule, and that there are clear grade differences between invertebrate taxa. Thus ants and web-building spiders appear to have similar lines that are to the left of the lines of reptiles, amphibians and teleost fish (Wehner *et al.*, 2007; Seid *et al.*, 2011, Fig. 1B), but to the right of the line for weevils. In insects, just as in vertebrates, different taxa have quite different brain–body allometries (Fig. 1B). In addition, the points for individual species of very small animals, including an immature crustacean and non-arthropod invertebrates such as kinorhynchs and nematodes suggest further grade changes, as they are substantially to the left of the lines of insects and spiders and their brains occupy only moderately small fractions of their bodies (Figs. 1B and 4).

How are some animals able to be so much smaller than others? Although it may be tempting to adopt a line of reasoning along the lines of the size limitation option, and suppose that animals in the ''lower'' grade are behaviourally inferior, such reasoning is unlikely to be consistently sustained by facts. Mammals and birds are indeed probably capable of more elaborate, diverse and adaptively flexible behaviour than are reptiles and salamanders. If such a general size-limitation type of idea is correct, however, one would expect that a honeybee should be behaviourally inferior to a miniature salamander, as it is in a grade whose line lies substantially to the left of that of the salamander (Fig. 1B). Although we lack a meaningful metric to quantify behavioural comparisons, we strongly doubt that this simple size-limitation prediction is true. Among other things, a honeybee can fly rapidly and manoeuvre accurately in the air, orient precisely over long distances, communicate complex messages to nestmates about the direction and distance of floral resources, make consensus-based decisions regarding new nest sites, master concepts such as sameness versus different and above versus below (Avargués-Weber *et al.*, 2011), build complex nests, learn diverse spatial and temporal patterns in its environment, and flexibly adjust foraging tactics to the array of resources that are available and to the needs of her colony (Seeley, 1985, 2010; von Frisch, 1967; see summary in Chittka and Niven, 2009).

A kinorhynch or a gastrotrich might be expected to be behaviourally inferior to a tiny beetle larva or a spider. At least some aspects of gastrotrich locomotory behaviour do seem relatively simple (though more complex than that of unicellular ciliates; Banchetti and Ricci, 1998) when compared, for example, with that of an orb-weaving spider; but their behaviour may not necessarily be simple when compared with that of a tiny ptiliid beetle larva, which has a comparatively gigantic brain (Fig. 4A). The larva probably feeds only on fungal spores it encounters while moving on or within the fruiting body of a fungus, where it lives from the time it hatches from an egg, and defends itself simply by crawling into a spore tube (W. Gene Hall, personal communication). The list of behaviours that a *C. elegans* nematode can perform with its tiny array of 302 structurally simple neurons is striking: it includes context-dependent release of neurosecretory products; sensitivity to various chemicals, O_2 concentration, gentle and harsh contact with other objects, osmolarity, pH, temperature, light and the density and sex of nearby conspecifics via pheromones; and motor functions that include different movements for swimming as well as for crawling on a surface; the ability to turn and to reverse these movements; pacemaker functions to impose rhythms on behaviours; orientation and movement toward and away from various stimuli; foraging for food; using rhythmic movements to engulf and swallow food; grinding ingested food; social feeding; defecation; copulation; oviposition; and learning to modify several of these motor behaviours on the basis of previous experience (Altun and Hall, 2008). Responses to a given stimulus can be complex; for instance, gentle contact can elicit avoidance, reset the defecation cycle, and suppress both pharyngeal movements and oviposition (Altun and Hall, 2008). One is reminded that studies of artificial nerve networks have shown that relatively few neurons are sufficient to produce surprisingly refined behaviour (Smith *et al.*, 2008); perhaps the most pertinent question is not how small brains produce complex behaviour, but rather what is the advantage of having large brains (Chittka and Niven, 2009).

Just as our understanding of the sensory and nervous systems of tiny marine and freshwater animals is based on very limited sampling, our understanding of their behaviour is even more fragmentary, especially for even smaller immature stages. Despite their relatively simple nervous systems (García-Arrarás *et al.*, 2001), larval echinoderms gather numerous particles with their cilia, but when these potential food items arrive at the mouth they may be either swallowed or rejected (Strathmann, 1975). In general, the larvae of many marine invertebrates express behaviour that is more sophisticated than generally appreciated, and, depending on the taxa, may involve sensors that enable functional responses to variation in water chemistry from biotic and abiotic sources; water-borne sounds; white light gradients and polarized light; water current movement and water pressure; and magnetism (reviewed in Kingsford *et al.*, 2002). For presettlement stages of the larvae of 11 marine taxa, there is a positive association between an index of sensory capabilities and swimming speed, such that faster (and presumably stronger) swimmers have more sensory inputs (Kingsford *et al.*, 2002).

In some taxonomic groups the properties of neurons and their connections may be specially designed in especially small individuals so as to allow maximal behavioural capabilities from minimal amounts of nervous tissue (the economy of design option). We give a speculative list of traits of possible interest in this context. The CNS of *C. elegans* displays several apparent economical traits, including muscle plate synapses that allow several muscles to be innervated by a single neuron (Fig. 2; also seen in a copepod nauplius—Lacalli, 2009), and widespread multifunctionality for particular neurons, including various combinations of sensory, motor and inter-neuron functions in single neurons, as well as diversity in a single modality such as sensitivity to several types of stimuli (Altun and Hall, 2008). Neurons with multiple functions also occur in tiny male polychaete worms (Windhoffer and Westheide, 1988).

One set of traits that may be associated with a grade change is the anatomy of individual neurons. In the "neuron-profligate" vertebrates, the CNS is dominated by one type of neuron morphology (heteropolar multipolar) that is found only here and there among invertebrates (especially in the visceral plexuses that resemble the diffuse ancestral coelenterate organization; Bullock and Horridge, 1965). Also in contrast with typical invertebrate neurons, many vertebrate synapses are on the cell body, rather than on the dendrites (Bullock and Horridge, 1965). Still another difference was recognized long ago by the pioneering neuroanatomist Ramon y Cajal (see the quote at the head of chapter): the neurons of insects are much more elaborate morphologically than those of vertebrates (see also Strausfeld, 1976). The functional consequences of these different morphologies are not known, so it is not clear whether any of these differences contribute to grade changes in brain allometry.

Morphological exhuberance in neuron morphology does not extend to some even smaller animals such nematodes like *C. elegans* and others, as their only modest numbers of neurons generally have simple processes that are mostly unbranched (Bullock and Horridge, 1965). Even the diameters of different fibres of a given neuron are relatively fixed (Bullock and Horridge, 1965), and the neurons make only very modest numbers of synaptic connections (*C. elegans* has a total of 6400 chemical and 900 gap junction synapses for 302 neurons, giving a mean of 24 synapses/cell; Altun and Hall, 2008). Some individual insect neurons have more branches than the total number of neurons in the entire nervous system of a nematode like *Ascaris* and *C. elegans* (Bullock and Horridge, 1965; Strausfeld, 1976). A less complete inventory of neurons in the nauplius larva of the tiny copepod *Dactylopusia* sp. suggests similarly small numbers and morphological simplicity. For instance, each retinular cell has a total of only ''one-several'' synaptic bars with the visual processing cells (Lacalli, 2009). Further studies are needed to understand whether other anatomical diversity within cells, such as the mysterious radial striations seen in many nematode neurons, their tight packing, lack of glial sheaths, and their apparently anastomosing processes (Bullock and Horridge, 1965), also have economy of design consequences.

Morphological differences associated with grade changes are also apparent at another level of organization. The internal division of the brains of very small insects into distinct subregions is less clear than in their larger relatives (Grebennikov, 2008); similarly, the brain of a copepod nauplius also lacks clear segmentation and connectives, although it does have distinct ''cartridges'' that receive visual inputs (their vision functions to trigger escape responses to shadows; Lacalli, 2009). In other extremely small marine invertebrate larvae (≤ 150 microns long), the nervous system is generally diffuse rather than condensed (Lacalli, 2009), suggesting still another possible miniaturization tactic. The fact that vertebrates rely almost exclusively on heteropolar multipolar neuron form suggests that their brain anatomy and function are organized quite differently, with cell bodies and neuropil mixed intimately together (Bullock and Horridge, 1965). In addition, they have a greater morphological variety of glial cells than invertebrates (Bullock and Horridge, 1965). Perhaps the relative numbers of neurons dedicated to sensory (incoming information), motor (outgoing information) and inter-neuron functions (analysis and coordination) also affects economy of design. Sensory and motor neurons constitute the vast majority of the nervous system of *C. elegans* (Altun and Hall, 2008); only 14.2% of the neurons in the CNS of *Ascaris lumbricoides* do not have direct connections with either sense organs or muscles.

Still another suite of traits that might be associated with grade changes and miniaturization concern intra- and inter-cellular transfer of information. Insects appear to more often utilize graded depolarizations (as opposed to the more energetically expensive action potentials) to transmit impulses (Chittka and Niven, 2009). Very tiny marine invertebrates typically lack synapses completely (Lacalli, 2009). In the nauplius of a copepod, three of the nine retinular cells of the eye, as well as some neural processes that end on muscles, lack synapses. The use of analogue transmission is energetically more efficient (see Section 5.3), but the functional significance of the latter two differences is unclear.

In sum, there are many possible kinds of economies of design, but at the moment their possible importance in grade changes is still highly speculative.

7.4 GENERAL IMPORTANCE AND CONSEQUENCES OF BRAIN SCALING

Problems of CNS size and design associated with miniature size probably have a variety of ecological and evolutionary consequences. We briefly discuss a few, many of which are only speculative, to illustrate the wide range of possibilities, and to highlight questions that are rarely asked and are in need of study.

Due to the higher metabolic demands of nerve tissue, the over-sized brain option to solve miniaturization problems should increase the probability that smaller animals will fall into unfavourable energy balances, unless they have compensatory traits. Simply maintaining a CNS that constitutes $>75\%$ of the entire cephalothorax likely results in sizeable energy costs for small anapid spiders. The energy budget of a newly emerged spiderling of *Nephila clavipes* (0.7 mg) and its susceptibility to shortages of resources must differ substantially from those of a mature female (2000 mg). In addition, relatively large brains take up space that would otherwise be profitably utilized by other structures. For instance, the digestive cecae that are abundant in the cephalothorax of an adult *Phidippus clarus* are nearly completely missing from the cephalothorax of a second instar nymph (Hill, 1975), presumably reducing the speed or efficiency of digestion of food.

Behaviour could also be affected in subtle ways. For instance, there are several types of learning in insects, and long-term memory storage can be more costly energetically (Hoedjes *et al.*, 2011). In small individuals, in which energy balance problems are likely to be more acute, animals might be more likely to avoid using long-term memory. Both forming memories and erasing them consume energy (Burns *et al.*, 2011), and smaller individuals might opt to minimize the use of memory or learning in general. The organization of behaviour could also be modified by efficiency of design adjustments, using sub-routines to render a given task less demanding. This tactic is employed by designers of robots (Clark, 1998), and by predators such as salticid spiders, which first turn to face directly toward the prey before they attempt to leap onto it.

Another phenomenon possibly related to brain miniaturization involves possible gains and losses of chromosomes. Recent studies have shown that neuronal cells of humans and mice can lose (or gain) a chromosome during normal ontogeny, so that the CNS is a genetic mosaic (Rehen *et al.*, 2001; Yurov *et al.*, 2005). These aneuploid neurons are not confined to a specific brain region (Kingsbury *et al.*, 2005), but are integrated into brain circuitry, and can alter gene expression profiles (Kaushal *et al.*, 2003). To our knowledge, nothing is known about the frequency of aneuploid neurons in arthropods, or if they are more frequent in small animals in general. We speculate that a facultative reduction in genome size could be a design economy to reduce the relative nucleus size, and in turn reduce overall cell size. Limited evidence suggests that evolutionary lineages with relatively small genomes do not seem more prone to evolve miniature species, as miniaturization is especially common in frogs and salamanders (Hanken and Wake, 1993), which have unusually large genomes (Gregory, 2002). Nevertheless, reduction in chromosome numbers would be particularly advantageous in just this type of situation. It would be worth searching for facultative genome reductions in the neurons of especially small animals, for more compact packing of their chromatin (Grebennikov, 2008), or for evidence that genes needed for neural function are concentrated on certain chromosomes, thus allowing other chromosomes to be discarded from neurons.

Differences in general morphological designs within a grade may also be illuminated by understanding brain–body scaling problems. We have already mentioned that the ventrally bulging sterna that are characteristic of small spiders (at least in some groups—R. Madrigal, in preparation) apparently represent adjustments to house their relatively large brains. A search for similar body design changes that correlate with relatively large brain volumes in especially small individuals of other groups might be rewarding. To illustrate possible consequences with one possible example, the brains of especially small ants constitute a very large proportion (16%) of their body weight (Seid *et al.*, 2011). Does the need to house this large brain in the head capsule impose a limit on the sizes of the mandibular muscles, and thus on the possible life styles of especially small ant species?

Possible ecological and life-history changes that may be associated with the energetic consequences of the over-sized brain and the size-limited behaviour options are summarized, respectively, in Table 2A and B. For example, endowing young nymphs with behavioural capabilities comparable to those of adults presumably improves their abilities to forage successfully. The relatively larger size of the eggs of smaller spiders, in terms of the female's body size (Head, 1995), could thus be due at least in part to brain scaling problems, because the

TABLE 2

A sampler of possible ecological consequences of adopting different options to reduce brain scaling problems in small animals (young individuals, or adults of small species)

A. Adopt over-sized brain option
- Reduced capacity to survive in sites where energy resources are less predictable or more scarce (could be due to differences in seasonality, rainfall, fluctuations in prey populations, etc.)
- Reduced foraging efficiency due to reduced use of long-term memory (more energetically expensive)
- Reduced optimality of body design for ecological tasks (agility, rapidity, strength, coordination) due to structural modifications needed to house the relatively large brain
- Reduced clutch sizes, resulting from increased relative size of offspring (due to lower limit of size of ecologically viable offspring)

B. Adopt size-limited behaviour option
- Reduced ability to adaptive behavioural flexibility to respond to changes in biotic and abiotic environment
- Reduced effectiveness of foraging, due to
  - Reduced sensory information
  - Reduced ability to integrate sensory information
  - Reduced ability to learn
  - Inferior motor coordination

The objective here is to call attention to the wide range of possibilities, not to demonstrate proven correlations.

costs of supporting proportionally larger amounts of nervous tissue in newly emerged nymphs may impose a lower limit on egg size. Smaller species of mites also have smaller clutches of relatively larger eggs (Walter and Proctor, 1999). Grebennikov (2008) notes that many hymenopteran egg parasitoids may escape such a limit because their larvae hatch into a host environment that provides necessary nutrients for them to develop, allowing them to evolve smaller body sizes than non-parasitic species.

For tiny predators such as orb-building spiders, these costs of relatively large brains may be exacerbated if prey that are small enough to be captured are rare. Several types of prey were found, however, being fed upon by adults of the small anapid *A. simoni* in the field; most were small nematocerous flies (56%), and others included ants, collembolans and mites. So at least at this body size (about 0.8 mg) a variety of prey are potentially available (no data are available on the prey of small nymphs of this species).

Similar considerations regarding lower size limits on ecologically functional offspring are probably important in many other groups, because spiders are probably the norm rather than the exception in generating very small offspring from medium sized individuals and in having lower feasible size limits for their offspring. The lower size limits of different taxa are very different. The deformation of brains and the structures that house them in the forms near the lower end of each spectrum (tiny salamanders, tiny beetles, spiders, etc.) suggest that brain allometry problems may be important in establishing these lower body size limits. Presumably the CNS design traits that determine the brain–body allometry line for a given grade or taxon originally evolved in intermediate or large-sized ancestors (this is the case for spiders, judging by recent phylogenies—Wheeler and Hayashi, 1998; Coddington, 2005). The evolution of warm-bloodedness may have helped moderately sized birds and mammals move independently onto allometric lines above that of their reptilian ancestors. Within such a derived group, forms with reduced body sizes may then evolve later, but only down to the limits imposed by the new basic design. At this body size limit, CNS sizes begin to become unsustainably large. As the species of an evolutionary line approach such lower body size limits, selection for compensatory traits, such as reductions in behavioural abilities, or increased efficiency in the amount of nerve tissue needed to generate behavioural abilities is likely to become more intense. The fragmentary data summarized above on orb-weaving spiders indicate that reductions in behavioural abilities have not occurred, despite the relatively great burden imposed by their relatively large CNS. On the other hand, reductions in the relative sizes of portions of the nervous system dedicated to functions other than vision suggest that behavioural losses (which have not been detected) may have occurred in salamanders (Hanken and Wake, 1993).

The likely high intensity of natural selection favouring design economy in such species means that their neurophysiology and neuroanatomy are likely to be especially interesting, and worthy of study. Possibly the reduced internal differentiation into separate substructures of the brain (central body, corpora pedunculata, commissures, chiasmata) in the tiny larvae of strepsipterans and ptiliid beetles (Beutel *et al.*, 2005) results from economy of design, using neurons for multiple tasks (Chittka and Niven, 2009). Similarly a single neuronal pathway in a rotifer appears to connect both the lateral horns and dorsolateral apical receptors, suggesting that convergence of synaptic connections may be common in the afferent sensory systems of rotifers (Hochberg, 2009). Some other modifications also seem plausible. The tendency toward greater condensation or fusion of the CNS in especially tiny insects (Grebennikov, 2008) may reflect a more favourable balance between transmission distances for intra-ganglion stimuli as opposed to sensory and motor stimuli that enter and leave the ganglion. A similar explanation may account for the recurrent fusion of neighbouring ganglia in the ventral nerve cord of insects (Niven *et al.*, 2008; Wille, 1961). On the other hand, why is it that the nervous systems of tiny larvae of marine invertebrates are dispersed rather than centralized (Lacalli, 2009)?

Many of the ideas here are only speculations, but it seems likely that exciting discoveries are waiting to be made from studies of the nervous systems of miniature animals.

7.5 LIMITATIONS OF CURRENT DATA AND QUESTIONS FOR THE FUTURE

Discussing the evolution of ''the'' brain as a single entity can be seriously misleading. A brain is a mosaic of many interacting but to some extent independent or modular regions that can evolve semi-independently. Wide-ranging comparisons like those presented here can be helpful in understanding brain evolution, but are they complicated by the fact that different functions may be performed in different portions of the CNS in different taxa (Table 3). For

TABLE 3
Divisions of the CNS in different taxa that correspond roughly to each other with respect to function

Vertebrates	
Brain	Spinal cord
Insects	
Brain (supra- + sub-esophageal ganglia)	Ventral nerve cord (thoracic and abdominal ganglia)
Spiders	
Supra-esophageal ganglion	Sub-esophageal ganglion
Tiny invertebrates	
Cephalic ganglion	Other ganglia

instance, the "brain" of an insect corresponds to some extent to the supra-esophageal ganglion of a spider, but there is no certainty that the functions that are carried out in the brain of an insect are necessarily performed in the supra-esophageal rather than the sub-esophageal ganglion of a spider. Comparing brains or their equivalents in different groups in order to understand behavioural capabilities may to some extent involve comparing apples and oranges. One hopes that improved knowledge of the functions of different portions of the nervous systems of different animals will permit a more precise focus in future comparisons. The question of whether some functions are performed at different sites (brain vs. spinal cord, head ganglia vs. ventral nerve cord) in different groups can be difficult to untangle. For example, learning typically is thought to occur in the brain, but simple lessons (e.g. altering leg position to avoid noxious, repeated stimuli) can be learned by the ventral nerve cord (VNC) of a headless cockroach (Horridge, 1962; learning was faster and retention was longer, however, when insects had an intact head/brain—Chen *et al.*, 1970). Similar spinal cord learning is also known in mice and other vertebrates (e.g. Jindrich *et al.*, 2009). The question is open as to whether or not miniature animals shunt control of some functions from the brain to ganglia in the VNC or vice versa. Comparisons involving the entire central nervous system (CNS) would thus be preferable, especially when widely different taxa are involved. But even here interpretation can be complicated, as the sense organs of some groups can perform functions performed by the CNS in other groups (Franzl and Barth, 2009; Wehner, 1987).

Unfortunately, most published data on insects and vertebrates regarding the allometry of nervous systems concern measurements of the "brains" rather than of the entire CNS, so we have been obliged to use these data rather than CNS data in most places. Even within vertebrates, the percentage of the CNS dedicated to the spinal cord (rather than the brain) varies widely; in 31 species it ranged from 2.3% in a porpoise to 37.1% in a domestic chicken (MacLarnon, 1996). The allometric slope of spinal cord mass on body mass is lower than that for brain on body mass, and reconstruction of a large dinosaur suggests its spinal cord mass was even greater than that of its brain (Striedter, 2005). MacLarnon (1996, p. 81) concluded that "... there are significant differences in the size of the cord relative to body size" between vertebrate classes. In insects there is a dearth of data to assess how the total nervous system is distributed between the brain and the ventral nerve cord (VNC; Niven *et al.*, 2008; Wille, 1961). In a subterranean termite, *Reticulitermes flavipes* (Rhinotermitidae), the cross-sectional areas for the brain, and each ventral nerve cord ganglion (three in the thorax and nine in the abdomen) differed with respect to the total nervous system between for female alates (57%), soldiers (48%) and workers (42%) (Nuss *et al.*, 2008). As noted in the caption of Fig. 1, values for "brains" of spiders and mites, which include ganglia corresponding to those of the VNC of insects, are higher than would be appropriate for comparisons with most other groups. Extreme miniaturization, as well as fusion of different ganglia, may alter these relative proportions (Beutel *et al.*, 2005; Grebennikov, 2008; Niven, 2010).

As noted above, behavioural data on tiny species are extremely limited in both number and degree of subtlety. Our current ignorance is probably due largely to the theoretical difficulty of quantifying behavioural traits, the practical difficulty of studying the behaviour of tiny animals, and the biased idea that larger animals (e.g. adults as opposed to immatures) are more interesting or important. An especially large hole in current knowledge of the behaviour of tiny species is a nearly complete lack of data on their abilities to learn. Tiny parasitoid wasps are able to learn relatively simple lessons such as changes in attraction to different hosts on the basis of the host plants on which their host larva developed (Hanson and Gauld, 2006; Papaj and Lewis, 1993), and the nematode *C. elegans* can learn lessons associated with several different types of stimuli (Altun and Hall, 2008); but the comparative ease with which large and small species learn the same lessons remains to be investigated. Much more work will be needed to obtain a reasonable evaluation of how frequently the size-limited option has been used.

Another significant limitation to understanding brain-behaviour relations is the dearth of comparative data resulting from the collective blinders imposed by the ''model system'' approach to studies of both behaviour and morphology; approximately 75% of research efforts by neuroscientists involve brains of mice, rats and humans, an infinitessimal fraction of the nervous systems on the planet (Manger *et al.*, 2008, p. 1). Another limitation is that we have been consistently typological throughout this discussion because of the lack of data on intra-specific variation, even though we know that the brains of vertebrates and insects (and probably other groups) show both genetic variation and experience-dependent plasticity in size and cell number. There are differences of up to 100% in the numbers of neurons for at least some functions in different races of the same species, as well as substantial variation even within members of the same population (Burns *et al.*, 2009; Heisenberg *et al.*, 1995; Miklos, 1998). We have used a single value for the weight of each species, when obviously weights vary substantially even for a single individual. We used the gross weight, without discriminating lean body mass, which correlates more strongly with brain weight in some species (Schoenemann, 2004). Hopefully future studies will be more sophisticated in this respect and assess the functional significance of variation in such traits. We have interpreted graphs without attempting to correct for phylogenetic inertia, but believe that the relatively tight groupings of the members of most taxa (except insects—see Fig. 1) around the same line for that taxon indicates that such correlations would not likely change our conclusions.

A significant deficiency in our understanding of the neuroanatomy of miniature invertebrates is data on how neuronal cell size and neuron number (and hence neuronal density) scale with body size (for vertebrates, see e.g. Herculano-Houzel, 2010; Sarko *et al.*, 2009). The limited studies that present data on cell size or neuron number (e.g. Babu, 1975; Babu and Barth, 1984 on spiders) do not always specify the stereological methods used, so it is not clear if appropriate sampling methods were used to infer three-dimensional properties from two-dimensional samples (see Howard and Reed, 1998).

The short-comings sketched here can serve as a roadmap to outstanding questions for the future, which will provide the data needed to more fully understand the causes and consequences of brain miniaturization, and its limitations. The length of the list emphasizes the speculative nature of much of our discussion. In many cases we lack data to answer these questions because they are rarely asked. We need to understand the causal mechanism(s) underlying Haller's Rule (our only progress in the present paper has been to argue against some previous hypotheses). The extent to which these design challenges shape animal evolution is not clear, nor is it clear to what extent different solutions have been adopted by different lineages. Detailed neurophysiological, neuroanatomical and behavioural data are needed for a wide taxonomic range of animals and life-history stages. Several taxonomic groups might be especially promising for future studies. Especially tiny invertebrates such as kinorhynchs and planktonic crustaceans are of obvious interest. Ants are also interesting, because they seem to lie on an allometric line below those of some other insects (Fig. 1B). Even relatively large ant species (which should presumably be easier to study) could be interesting, as they lie on the same line. In addition, detailed behavioural studies of leaf-cutter ants, *Atta*, in which a huge range of sizes is spanned in a single species, checking tiny workers for signs of size limitation such as reduced ability to learn, greater clumsiness, higher error rates, or reduced behaviour repertoires could be interesting. Another interesting group is the mites. The fragmentary data currently available suggest that the allometric line of mites lies below those of insects and spiders (Fig. 1B). In accord with this, mites have evolved adult body sizes substantially below those of the smallest insects and spiders (Grebennikov, 2008). Some mites are nevertheless capable of analyses and learning, using a brain with only approximately 10,000 neurons (van Wijk *et al.*, 2006). Both the behaviour and the nervous systems of this poorly studied group would repay further study.

As we noted above, many brain-behaviour associations may be obscured by too much lumping. We do not know the extent to which the same behaviours in different lineages are governed by the same brain regions, or whether different lineages rely to different extents on mechanisms such as learning. For most taxa of small invertebrates we need stereologically valid (Howard and Reed, 1998) data on neuronal cell size, number and density to assess whether animals with extremely small body sizes maintain information-processing capabilities comparable to those of large-bodied species via a reduction in neuron size. Similarly, we do not know how taxa of strikingly different body sizes have adjusted the relative numbers of glia and non-glia cells, or more generally whether different taxa follow different cellular scaling rules (e.g. Herculano-Houzel *et al.*, 2006; Sarko *et al.*, 2009). Understanding behavioural variation requires a way to replicate in other taxa the kinds of data that have been collected with orb-weaving spiders, which in turn requires methods to make quantitative behavioural comparisons among disparate taxa.

7.6 CONCLUSIONS

The evolution of small body sizes poses a number of challenges to animals, because smaller individuals tend to have larger and disproportionately costly brains, which reach, in small insects and spiders, up to 15–16% of the animal's body mass. To accommodate these relatively large brains, some species use other portions of the body to house the CNS, deform structures to increase internal space, or displace internal tissues. This would imply that extremely small animals sacrifice some morphological aspects of body design to accommodate their relatively large brains. Their relatively large brains are also probably costly, because nervous tissue is metabolically expensive to maintain. These problems of diminutive size are of very general importance, because they occur not only in species with miniature adults (and even smaller immatures), but also in many other species with large adults and with very small, free-living early developmental stages.

Increased intensity of natural selection favouring reduced costs in small animals could favour traits that are less demanding in terms of behavioural capabilities. To date, however, the only detailed behavioural studies that are available, concerning the construction behaviour of orb web spiders, suggest that very small animals are not behaviourally impaired compared with larger individuals.

The brain scaling in insects and spiders is quantitatively different from that of vertebrates, and probably also from that of very small invertebrates such as nematodes, tiny crustaceans, gastrotrichs and kinorhynchs. Such ''grade changes'' in brain allometry are probably not always linked in simple ways to differences in behavioural capabilities, and may instead (or in addition) be due to differences in the anatomy of their neurons and other basic aspects of how their nervous systems operate. We have provided largely speculative lists of possible economies in design. The idea that there are profound differences in mechanisms of brain function in different taxa seems to fly in the face of some basic similarities in the anatomical designs of sets of neurons subserving similar sensory processes in widely different taxa (e.g. structures analysing olfactory stimuli in insects, crustaceans and vertebrates; visual pathways in cephalopods, vertebrates and arthropods—Strausfeld, 1976; Chittka and Niven, 2009). Nevertheless, the contrast between the few hundred structurally simple neurons whose modest numbers of processes are connected computer-like, only in certain, consistent ways in the brains of nematodes (Fig. 14; Altun and Hall, 2008; Bullock and Horridge, 1965), as opposed to the astronomical numbers of neurons in a human brain, with huge numbers of connections for each cell (a typical pyramidal cell in the human cortex can have 10,000 connections—Edelman, 1998), whose numbers, properties and connections

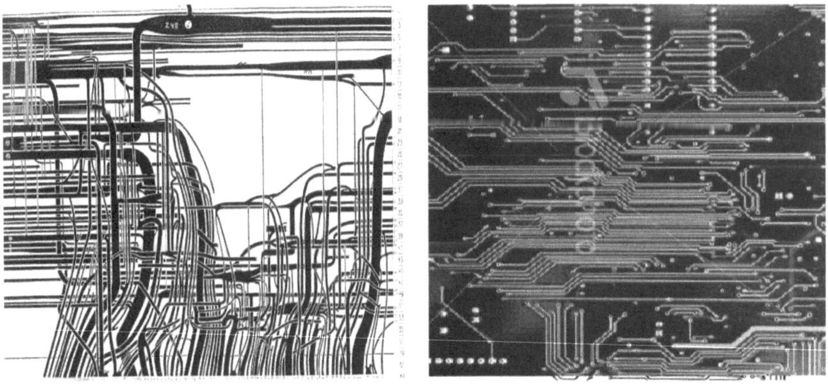

Ascaris megalocephala Goldschmidt 1909

Chip computer Google Images 2011

FIG. 14 Diagrammatic representation (left) of the connections of 70 of the 134 fibres in the circumenteric ring near the origin of the ventral nerve cord of the nematode *Ascaris megalocephala* (from Bullock and Horridge, 1965, after Goldschmidt, 1909). The uncanny visual resemblance to a computer chip (right) emphasizes the computer-like traits of nematode nervous systems (fixed numbers of elements, invariant connections between them) that are exactly the computer-like aspects that do *not* occur in vertebrate nervous systems (Edelman, 1998). This emphasizes the likely profound differences in how the brains of taxa in different grades function. (See Color Plate 4 at the back of the book.)

are continually shifting with use and disuse, whose patterns of behaviour are based only on population trends rather than individual consistency, and which show a highly degenerate organization that provides multiple physical representations of something like a memory (Edelman, 1998), suggests quite different modes of operation. We believe that grade changes, which we have emphasized in this paper, have probably involved the evolution of such basic differences.

Acknowledgements

We thank Kevin Carman, Thursten Lacalli, Rachel Collin, John Christy, Rick Hochberg, Jeremy Niven and Heather Proctor for help finding literature and other advice. We thank John Douglass, Ruth Madrigal, Gloria Vargas, and Marc Seid for permission to use unpublished data, and the F. H. Levinson Family Foundation (W. T. W), the Smithsonian Institution Scholarly Studies Program (W. T. W.), the Universidad de Costa Rica (W. G. E) and the Smithsonian Tropical Research Institute (W.G. E and W. T. W) for financial support. We also thank Sabrina Amador, Mary Jane West-Eberhard, Jeremy Niven, and Andre Riveros, for kindly providing helpful comments on an earlier version, and Jerome Casas for the invitation to write this paper.

References

Altun, Z. F. and Hall, D. H. (2008). Handbook of *C. elegans* anatomy. *In* WormAtlas http://www.wormatlas.org/hermaphrodite/hermaphroditehomepage.htm.

Anderson, M. L. (2010). Neural reuse: a fundamental organizational principle of the brain. *Behav. Brain Sci.* **33**, 245–313.

Anderson, A. and McShea, D. W. (2001). Individual versus social complexity, with particular reference to ant colonies. *Biol. Rev.* **76**, 211–237.

Avargués-Weber, A., Dyer, A. G. and Giurfa, M. (2011). Conceptualization of above and below relationships by an insect. *Proc. R. Soc. B* **278**, 898–905.

Babu, K. S. (1975). Post embryonic development of the central nervous system of the spider *Argiope aurantia* (Lucas). *J. Morphol.* **146**, 325–342.

Babu, K. S. and Barth, F. (1984). Neuroanatomy of the central nervous system of the wandering spider, *Cupiennius salei* (Arachnida, Araneida). *Zoomorphology* **104**, 344–359.

Banchetti, R. and Ricci, N. (1998). The behavior of *Heterolepidoderma* sp. (Gastrotricha). *Zool. Sci.* **15**, 131–137.

Barnes, R. S. K., Callow, P. and Olive, P. J. W. (1993). The Invertebrates: A New Synthesis. Blackwell Scientific Publications, Boston, MA.

Barth, F. G. (2002). A Spider's World: Senses and Behavior. Springer Verlag, Berlin.

Bernays, E. A. and Wcislo, W. T. (1994). Sensory capabilities, information processing, and resource specialization. *Q. Rev. Biol.* **69**, 187–204.

Beutel, R. G. and Haas, A. (1998). Larval head morphology of *Hydroscapha natans* LeConte 1874 (Coleoptera, Myxophaga, Hydroscaphidae) with special reference to miniaturization. *Zoomorphology* **18**, 103–116.

Beutel, R. G., Pohl, H. and Hunefeld, F. (2005). Strepsipteran brains and effects of miniaturization (Insecta). *Arthropod Struct. Dev.* **34**, 301–313.

Bonner, J. (2006). Why Size Matters. Princeton University Press, Princeton, NJ.

Brodin, A. (2005). Hippocampus volume does not correlate to food hoarding rates in two closely related bird species, the black-capped chickadee *Poecile atricapilla* and the willow tit *Parus montanus*. *Auk* **122**, 819–828.

Brown, F. A. (1950). Selected Invertebrate Types. John Wiley and Sons, New York, NY.

Brusca, R. C. and Brusca, G. J. (1990). Invertebrates. Sinauer, Sunderland, MA.

Bullock, T. H. and Horridge, G. A. (1965). Structure and Function in the Nervous Systems of Invertebrates W. H. Freeman and Co, San Francisco, CA(Vols. I, II).

Burger, J. M. S., Huang, Y., Hemerik, L., van Lenteren, J. C. and Vet, L. E. M. (2006). Flexible use of patch-leaving decisions in a parasitoid wasp. *J. Insect Behav.* **19**, 155–170.

Burns, J. G., Arthy Saravanan, A. and Rodd, F. H. (2009). Rearing environment affects the brain size of guppies: lab-reared guppies have smaller brains than wild-caught guppies. *Ethology* **115**, 122–133.

Burns, J. G., Foucaud, J. and Mery, F. (2011). Costs of memory: lessons from 'mini' brains. *Proc. R. Soc. B.* **278**.

Byrne, R. W. (1997). Machiavellian intelligence. *Evol. Anthropol.* **5**, 172–180.

Calder, W. A. I. I. I. (1984). Size, Function, and Life History. Harvard University Press, Cambridge, MA.

Calvin, W. H. and Stephens, C. F. (1967). Synaptic noise as a source of variability in the interval between action potentials. *Science* **155**, 842–844.

Calvin, W. H. and Stephens, C. F. (1968). Synaptic noise and other sources of randomness in motoneuron interspike intervals. *J. Neurophysiol.* **31**, 574–587.

Campi, K. L. and Krubitzer, L. (2010). Comparative studies of diurnal and nocturnal rodents: differences in lifestyle result in alterations in cortical field size and number. *J. Comp. Neurol.* **518**, 4491–4512.

Changizi, M. A. (2003). Relationship between number of muscles, behavioral repertoire size, and encephalization in mammals. *J. Theor. Biol.* **220**, 157–168.
Chapman, R. F. (1982). Chemoreception: the significance of receptor numbers. *Adv. Insect Physiol.* **16**, 247–356.
Chen, W. Y., Aranda, L. C. and Luco, J. V. (1970). Learning and long- and short-term memory in cockroaches. *Anim. Behav.* **18**, 725–732.
Cheng, K., Srinivasan, M. V. and Zhang, S. W. (1999). Error is proportional to distance measured by honeybees: Weber's law in the odometer. *Anim. Cogn.* **2**, 11–16.
Cherniak, C. (1995). Neural component placement. *Trends Neurosci.* **18**, 522–527.
Cherniak, C., Mokhtarzada, Z., Rodriguez-Esteban, R. and Changzi, K. (2004). Global optimization of cerebral cortex layout. *Proc. Natl. Acad. Sci. USA* **101**, 1081–1086.
Chittka, L. and Niven, J. (2009). Are bigger brains better? *Curr. Biol.* **19**, 995–1008.
Clark, A. (1998). Where brain, body and world collide. *Daedalus* **127**, 257–280.
Clark, B. and Häusser, M. (2006). Neural coding: hybrid analog and digital signalling in axons. *Curr. Biol.* **16**, R585–R588.
Coddington, J. A. (1983). The weavers. *Discovery* **4**, 48–53.
Coddington, J. A. (1986a). The monophyletic origin of the orb web. In: *Spider Webs Behavior and Evolution* (ed Shear, W.), pp. 319–363. Stanford University Press, Palo Alto, CA.
Coddington, J. A. (1986b). The genera of the spider family Theridiosomatidae. *Smiths. Contrib. Zool.* **422**, 1–96.
Coddington, J. A. (2005). Phylogeny and classification of spiders. In: *Spiders of North America* (eds Ubick, D., Paquin, P., Cushing, P. E. and Roth, V.), American Arachnological Society, www.americanarachnology.org.
Coddington, J. A., Hormiga, G. and Scharff, N. (1997). Giant female or dwarf male spiders? *Nature* **385**, 687–688.
Cole, B. J. (1985). Size and behavior in ants: constraints on complexity. *Proc. Natl. Acad. Sci. USA* **82**, 8548–8551.
Collett, T. S. and Collett, M. (2002). Memory use in insect visual navigation. *Nat. Rev. Neurosci.* **3**, 542–552.
Comstock, J. H. (1967). The Spider Book. Cornell Univ. Press/Comstock Publ. Assoc, Ithaca, NY.
Coyle, F. A. (1971). Systematics and natural history of the mygalomorph spider genus *Antrodiaetus* and related genera (Araneae: Antrodiaetidae). *Bull. Mus. Comp. Zool.* **141**, 269–402.
Coyne, J. A. (2009). Why Evolution is True. Viking Press, New York, NY.
Douglass, J. K., Kamhi, J. F. and Wcislo, W. T. (in preparation). Insect brain allometric relationships: A comparative study providing insights into space allocation trade-offs.
Dusenbery, D. B. (1992). Sensory Ecology. W. H. Freeman & Co, New York, NY.
Eberhard, W. G. (1969). The spider Uloborus diversus and its web. (PhD. Thesis, Harvard University).
Eberhard, W. G. (1972). The web of *Uloborus diversus* (Aranea: Uloboridae). *J. Zool. Lond.* **166**, 417–465.
Eberhard, W. G. (1982). Behavioral characters for the higher classification of orb-weaving spiders. *Evolution* **36**, 1067–1095.
Eberhard, W. G. (1986). Effects of orb-web geometry on prey interception and retention. In: *Spider Webs Behavior and Evolution* (ed Shear, W.), pp. 70–100. Stanford University Press, Palo Alto, CA.
Eberhard, W. G. (1987a). The effect of gravity on temporary spiral construction by the spider *Leucauge mariana* (Araneae: Tetragnathidae). *J. Ethol.* **5**, 29–36.
Eberhard, W. G. (1987b). Orb webs and construction behavior in Anapidae, Symphytognathidae, and Mysmenidae. *J. Arachnol.* **14**, 339–356.

Eberhard, W. G. (1988a). Behavioral flexibility in orb web construction: effects of silk supply in different glands. *J. Arachnol.* **16**, 303–320.

Eberhard, W. G. (1988b). Memory of distances and directions moved as cues during temporary spiral construction in the spider *Leucauge mariana* (Araneae: Araneidae). *J. Ins. Behav.* **1**, 51–66.

Eberhard, W. G. (1990a). Early stages of orb construction by *Philoponella, Leucauge,* and *Nephila* spiders (Araneae: Uloboridae and Araneidae). *J. Arachnol.* **18**, 205–2340.

Eberhard, W. G. (1990b). Imprecision in the behavior of larval fungus gnats and the origin of new behavior patterns in animals. *J. Ins. Behav.* **3**, 327–357.

Eberhard, W. G. (1990c). Niche expansion in the spider *Wendilgarda galapagensis* (Araneae, Theridiosomatidae) on Cocos Island. *Rev. Biol. Trop.* **37**, 163–168.

Eberhard, W. G. (2000). Breaking the mold: behavioral imprecision and evolutionary innovation in *Wendilgarda* spiders (Araneae, Theridiosomatidae). *Ethol. Ecol. Evol.* **12**, 223–235.

Eberhard, W. G. (2007a). Miniaturized orb weaving spiders: behavioral precision is not limited by small size. *Proc. Roy. Soc. Lond B* **274**, 2203–2209.

Eberhard, W. G. (2007b). Recovery of spiders from the effects of parasitic wasps: implications for fine-tuned mechanisms of manipulation. *Anim. Behav.* **79**, 375–383.

Eberhard, W. G. (2011). Are smaller animals behaviorally limited? Lack of clear constraints in miniature spiders. *Anim. Behav.* **81**, 813–823.

Eberhard, W. G. and Barrantes, G. (in preparation). A new behavioral test with Zosis geniculatus (Araneae, Uloboridae) supports orb web monophyly.

Eberhard, W. G. and Hesselberg, T. (submitted). Errors by spiders (Araneae: Araneidae, Tetragnathidae) building orbs: lapses in attention due to recent memories? *Ethology.*

Edelman, G. (1998). Building a picture of the brain. *Daedalus* **127**(2), 37–70.

Edmunds, J. (1978). The web of *Paraneus cyrtoscapus* (Pocock, 1899) (Araneae: Araneidae) in Ghana. In: *Arachnology, Seventh International Congress* (ed Merrett, P.), vol 42, , p. 485. Academic Press, New York, NY.

Eisenberg, J. G. (1981). The mammalian radiations. *An analysis of Trends in Evolution, Adaptation and Behavior.* University of Chicago Press, Chicago, IL.

Eisenberg, J. F. and Wilson, D. E. (1978). Relative brain size and feeding strategies in the Chiroptera. *Evolution* **32**, 740–751.

Faisal, A. A., White, J. A. and Laughlin, S. B. (2005). Ion-channel noise places limits on the miniaturization of the brain's wiring. *Curr. Biol.* **15**, 1143–1149.

Fankhauser, G., Vernon, J. A., Frank, W. H. and Slack, W. V. (1955). Effect of size and number of brain cells on learning in larvae of the salamander, *Triturus viridescens. Science* **122**, 692–693.

Felsenstein, J. (1985). Phylogenies and the comparative method. *Am. Nat.* **125**, 1–15.

Foelix, R. F. (1996). Biology of Spiders. 2nd ed. Harvard University Press, Cambridge, MA.

Franzl, P. and Barth, F. G. (2009). Biomaterial systems for mechanosensing and actuation. *Nature* **462**, 442–448.

García-Arrarás, J. E., Rojas-Soto, M., Jiménez, L. B. and Díaz-Miranda, L. (2001). The enteric nervous system of echinoderms: unexpected complexity revealed by neurochemical analyses. *J. Exp. Biol.* **204Ñ**, 865–873.

Gaston, K. J., Chown, S. L. and Mercer, R. D. (2001). The animal species-body size distribution of Marion Island. *Proc. Natl. Acad. Sci. USA* **98**, 14493–14496.

Goldschmidt, R. (1909). Das Nervensystem von *Ascaris lumbricoides* und *megalocephala* II. *Z. wiss. Zool.* **92**, 306–357.

Gonzalez-Voyer, A., Winberg, S. and Kolm, N. (2009). Distinct evolutionary patterns of brain and body size during adaptive radiation. *Evolution* **63**, 2266–2274.

Grebennikov, V. V. (2008). How small you can go: factors limiting body miniaturization in winged insects with a review of the pantropical genus *Discheramocephalus* and description of six new species of the smallest beetles (Pterygota: Coleoptera: Ptiliidae). *Eur. J. Ent.* **105**, 313–328.

Grebennikov, V. V. and Beutel, R. G. (2002). Morphology of the minute larva of *Ptinella tenella*, with specific reference to effects of miniaturization and the systematic position of Ptiliidae (Coleoptera: Staphylinoidea). *Arthrop. Struct. Dev.* **31**, 157–172.

Gregory, T. R. (2002). A bird's-eye view of the C-value enigma: genome size, cell size, and metabolic rate in the class Aves. *Evolution* **56**, 121–130.

Griswold, C. E., Coddington, J. A., Hormiga, G. and Scharff, N. (1998). Phylogeny of the orb-building spiders (Araneae, Orbiculariae: Deinopidae, Araneoidea). *Zool. J. Linn. Soc.* **123**, 1–99.

Gronenberg, W. and Riveros, A. J. (2009). Social brains and behavior-past and present. In: *Organization of Insect Societies* (eds Gadau, J. and Fewell, J.), pp. 377–401. Harvard University Press, Cambridge.

Hanken, J. (1983). Miniaturization and its effects on cranial morphology in plethodontid salamanders, genus *Thorius* (Amphibia: Plethodontidae) II. The fate of the brain and sense organs and their role in skull morpholgenesis and evolution. *J. Morphol.* **177**, 255–268.

Hanken, J. and Wake, D. B. (1993). Miniaturization of body size: organismal consequences and evolutionary significance. *Ann. Rev. Ecol. Syst.* **24**, 501–519.

Hanson, P. E. and Gauld, I. D. (2006). 2.5 Los parasitoides: la hembra adulta. In: *Hymenoptera del Neotropico* (eds Hanson, P. E. and Gauld, I. D.), **77**, pp. 49–62, (1–994).

Harvey, P. H. and Pagel, M. D. (1991). The Comparative Method in Evolutionary Biology. Oxford University Press, New York, NY.

Harvey, P. H., Clutton-Brock, T. H. and Mace, G. M. (1980). Brain size and ecology in small mammals and primates. *Proc. Natl. Acad. Sci. USA* **77**, 4387–4389.

Hasenstaub, A., Otte, S., Callaway, E. and Sejnowski, T. J. (2010). Metabolic cost as a unifying principle governing neuronal biophysics. *Proc. Natl. Acad. Sci. USA* **107**, 12329–12334.

Head, G. (1995). Selection on fecundity and variation in the degree of sexual size dimorphism among spider species (Class Araneae). *Evolution* **49**, 776–781.

Healy, S. D. and Rowe, C. (2007). A critique of comparative studies of brain size. *Proc. R. Soc. B* **274**, 453–464.

Heisenberg, M., Heusipp, M. and Wanke, C. (1995). Structural plasticity in the *Drosophila* brain. *J. Neurosci.* **15**, 1951–1960.

Herculano-Houzel, S. (2010). Coordinated scaling of cortical and cerebellar numbers of neurons. *Front. Neuroanat.* **4**, 1–8.

Herculano-Houzel, S., Mota, B. and Lent, R. (2006). Cellular scaling rules for rodent brains. *Proc. Natl. Acad. Sci. USA* **103**, 12138–12143.

Hesselberg, T. (2010). Ontogenetic changes in web design in two orb-web spiders. *Ethology* **116**, 535–545.

Higgins, R. P. (1983). The Atlantic barrier reef ecosystem at Carrie Bow Cay, Belize. II. Kinorhyncha. *Smiths. Contrib. Mar. Sci.* **18**, 1–131.

Hill, D. E. (1975). The structure of the central nervous system of jumping spiders of the genus *Phidippus* (Araneae: Salticidae). (M.S. thesis. Oregon State University. pp. 1–94).

Hingston, R. W. G. (1920). *A Naturalist in the Himalaya*. Small, Maynard and Co, Boston, MA.

Hochberg, R. (2009). Three-dimensional reconstruction and neural map of the serotonergic brain of *Asplanchna brightwellii* (Rotifera, Monogononta). *J. Morphol.* **270**, 430–441.

Hoedjes, K. M., Kruidhof, H. M., Huigens, M. E., Dicke, M., Vet, L. E. M. and Smid, H. M. (2011). Natural variation in learning rate and memory dynamics in parasitoid wasps: opportunities for converging ecology and neuroscience. *Proc. R. Soc. B* **278**, 889–897.

Homberg, U. (2008). Evolution of the central complex in the arthropod brain with respect to the visual system. *Arthropod Struct. Dev.* **37**, 347–362.

Homberg, U., Heinze, S., Pfeiffer, K., Kinoshita, M. and el Jundi, B. (2011). Central neural coding of sky polarization in insects. *Phil. Trans. R. Soc. Lond. B Biol. Sci.* **366**, 680–687.

Hormiga, G., Scharff, N. and Coddington, J. A. (2000). The phylogenetic basis of sexual size dimorphism in orb-weaving spiders (Araneae, Orbiculariae). *Syst. Biol.* **49**, 435–462.

Horridge, G. A. (1962). Learning of leg position by the ventral nerve cord in headless insects. *Proc. R. Soc. Lond. B* **157**, 33–52.

Howard, C. V. and Reed, M. G. (1998). Unbiased Stereology: Three-dimensional Measurement in Microscopy. Bios Scientific Publishers, Oxford.

Howse, P. E. (1974). Design and function in the insect brain. In: *Experimental Analysis of Insect Behaviour* (ed Barton-Browne, L.), Springer, New York.

Iwaniuk, A. N. and Wylie, D. R. W. (2007). Neural specialization for hovering in hummingbirds: hypertrophy of the pretectal nucleus lentiformis mesencephali. *J. Comp. Neurol.* **500**, 211–221.

Iwaniuk, A. N., Nelson, J. E. and Whishaw, I. Q. (2000). The relationships between brain regions and forelimb dexterity in marsupials (Marsupialia): a comparative test of the principle of proper mass. *Aust. J. Zoo.* **48**, 99–110.

Iwata, K. and Sakagami, S. F. (1966). Gigantism and dwarfism in bee eggs in relation to the modes of life, with notes on the number of ovarioles. *Jpn. J. Ethol.* **16**, 4–16.

Jander, U. and Jander, R. (2002). Allometry and resolution of bee eyes (Apoidea). *Arthropod Struct. Dev.* **30**, 179–193.

Jerison, H. J. (1973). Evolution of the Brain and Intelligence. Academic Press, New York.

Jerison, H. J. (1983). The evolution of the mammalian brain as an information-processing system. In: *Advances in the Study of Mammalian Behavior* (eds Eisenberg, J. F. and Kleinman, D. G.), The American Society of Mammalogist, New York.

Jindrich, D. L., Joseph, M. S., Otoshi, C. K., Wei, R. Y., Zhong, H., Roy, R. R., Tillakaratne, N. J. K. and Edgerton, V. R. (2009). Spinal learning in the adult mouse using the Horridge paradigm. *J. Neurosci. Methods* **182**, 250–254.

Kaestner, A. (1968). Invertebrate Zoology. Interscience Publishers, New York, NY, Vol. II. (Translated and adapted by H. W. Levi and L. Levi).

Katz, P. (1999). Beyond Neurotransmission: Neuromodulation and Its Importance for Information Processing. Oxford University Press, Oxford.

Kaushal, D., Contos, J. J. A., Treunere, K., Yang, A. H., Kingsbury, M. A., Rehen, S. K., McConnell, M. J., Okabe, M., Barlow, C. and Chun, J. (2003). Alteration of gene expression by chromosome loss in the postnatal mouse brain. *J. Neurosci.* **23**, 5599–5606.

Kelber, A., Warrant, E., Pfaff, M., Wallén, R., Theobald, J. C., Wcislo, W. T. and Raguso, R. A. (2006). Light intensity limits foraging activity in nocturnal and crepuscular bees. *Behav. Ecol.* **17**, 63–72.

Keshmirian, J. and Nogrady, T. (1987). Histofluorescent labeling of catecholaminergic structures in rotifers (Aschelminthes) in whole animals. *Histochemistry* **87**, 351–357.

Kiltie, R. A. (2000). Scaling of visual acuity with body size in mammals and birds. *Funct. Ecol.* **14**, 226–234.
Kingsbury, M. A., Friedman, B., McConnell, J. J., Rehen, S. K., Yang, A. H., Kaushal, D. and Chun, J. (2005). Aneuploid neurons are functionally active and integrated into brain circuitry. *Proc. Natl. Acad. Sci. USA* **102**, 6143–6147.
Kingsford, M. J., Leis, J. M., Shanks, A., Lindeman, K. C., Morgan, S. G. and Pineda, J. (2002). Sensory environmental, larval abilities and local self-recruitment. *Bull. Mar. Sci.* **70**, 309–340.
Kollmann, M., Bonhomme, J., Homberg, U., Schachter, J., Tagu, D. and Anton, S. (2011). Revisiting the anatomy of the central nervous system of a hemimetabolous model insect species: the pea aphid Acyrthosiphon pisum. *Cell Tissue Res.* **343**, 343–355.
Krink, T. and Vollrath, F. (1999). A virtual robot to model the use of regenerated legs in a web-building spider. *Anim. Behav.* **57**, 223–232.
Kristensen, R. M. and Higgins, R. P. (1984). Revision of *Styraconyx* (Tardigrada: Halechiniscidae), with description of two species from Disko Bay, West Greenland. *Smithson. Contrib. Zool.* **391**(iii-40).
Kuntner, M., Coddington, J. A. and Hormiga, G. (2008). Phylogeny of extant nephilid orb-weaving spiders (Araneae, Nephilidae): testing morphological and ethological homologies. *Cladistics* **24**, 147–217.
Lacalli, T. C. (2009). Serial EM analysis of a copepod larval nervous system: naupliar eye, optic circuitry, and prospects for full CNS reconstruction. *Arthrop. Struct. Dev.* **38**, 361–375.
Land, M. F. and Nilsson, D. E. (2002). Animal Eyes. Oxford Animal Biology Series, Oxford.
Lande, R. (1979). Quantitative genetic analysis of multivariate evolution, applied to brain–body size allometry. *Evolution* **33**, 402–416.
LeGuelte, L. (1967). Structure de la toile de Zygiella x-notata Cl. et facterus que regissent le comportement de l'araignee pendant la construction de la toile. (PhD. Thesis, Univ. Nancy, France).
Lopardo, L., Giribet, G. and Hormiga, G. (2010). Morphology to the rescue: molecular data and the signal of morphological characters in combined phylogenetic analyses—a case study from mysmenid spiders (Araneae, Mysmenidae), with comments on the evolution of web architecture. *Cladistics* **27**, 278–330.
MacLarnon, A. (1996). The scaling of gross dimensions of the spinal cord in primates and other species. *J. Human Evol.* **30**, 71–87.
Madrigal-Brenes, R. (in preparation). Sternal bulges in spiders of different sizes.
Main, B. Y. (1976). Spiders. Collins, London.
Manger, P. R., Cort, J., Ebrahim, N., Goodman, A., Henning, J., Karolia, M., Rodrigues, S.-L. and Štrkalj, G. (2008). Is 21st century neuroscience too focussed on the rat/mouse model of brain function and dysfunction? *Front. Neuroanat.* **2**, 1–7.
Mares, S., Ash, L. and Gronenberg, W. (2005). Brain allometry in bumblebee and honey bee workers. *Brain Behav. Evol.* **66**, 50–61.
Martin, R. D. (1981). Relative brain size and basal metabolic-rate in terrestrial vertebrates. *Nature* **293**, 57–60.
Martini, E. (1916). Die Anatomie der *Oxyuris curvula. Z. wiss. Zool.* **116**, 137–534.
Mayer, G. (1952). Untersuchengen ueber Herstellung und Struktur des Radnetz von Aranea diadema und Zilla x-notata mit besonderer Beruecksichtigung des Unterschiedes von Jungend- und Altersnetzen. *Z. Tierpsychol.* **9**, 337–362.
Mercer, A. (1999). Changing the way we perceive things: sensory systems modulation. In: *Beyond Neurotransmission: Neuromodulation And Its Importance for Information Processing* (ed Katz, P.), pp. 198–240. Oxford University Press, Oxford.

Michener, C. D. (2001). Comments on minute Meliponini and the male of the genus *Paratrigona* (Hymenoptera: Apidae). *J. Kansas Entomol. Soc.* **74**, 231–236.

Miklos, G. L. G. (1998). The evolution and modification of brains and sensory systems. *Daedalus* **127**, 197–216.

Miller, G. A., Galanter, E. and Pribram, K. H. (1960). *Plans and the Structure of Behavior*. Henry Holt & Co., NY.

Misunami, M., Yokohari, F. and Takahata, M. (2004). Further exploration into the adaptive design of the arthropod "microbrain": I. Sensory and mamory-processing systems. *Zool. Sci.* **21**, 1141–1151.

Niven, J. E. (2010). Nervous system evolution in relation to behavior. In: *Encyclopedia of Animal Behavior, Vol. 2* (eds Breed, M. D. and Moore, J.), pp. 527–533. Oxford Academic Press, New York, NY.

Niven, J. E. and Laughlin, S. B. (2008). Energy limitation as a selective pressure on the evolution of sensory systems. *J. Exp. Biol.* **211**, 1792–1804.

Niven, J. E., Anderson, J. C. and Laughlin, S. B. (2007). Energy-information trade-offs during signal transmission by single sensory receptors. *PLoS Biol.* **5**, e116.

Niven, J. E., Graham, C. M. and Burrows, M. (2008). Diversity and evolution of the insect ventral nerve cord. *Annu. Rev. Entomol.* **53**, 253–271.

Nuss, A. B., Forschler, B. T., Crim, J. W. and Brown, M. R. (2008). Distribution of neuropeptide F-like immunoreactivity in the eastern subterranean termite, *Reticulitermes flavipes*. *J. Insect Sci.* **8**, 1536–2442.

Osborne, K. A., de Belle, J. S. and Sokolowski, M. B. (2001). Foraging behaviour in *Drosophila* larvae: mushroom body ablation. *Chem. Senses* **26**, 223–230.

Papaj, D. R. and Lewis, A. C. (eds.) (1993). In: *Insect Learning*, Chapman & Hall, NY.

Platnick, N. and Shadab, M. (1978). A review of the spider genus *Anapis* (Araneae: Anapidae) with a dual cladistic analysis. *Am. Mus. Novit.* **2663**, 1–23.

Platnick, N. I. and Shadab, M. (1979). A review of the spider genera *Anapisona* and *Pseudanapis* (Araneae, Anapidae). *Am. Mus. Novit.* **2672**, 1–20.

Polilov, A. A. (2005). Anatomy of the feather-winged beetles *Acrotrichis montandoni* and *Ptilium myrmecophilum* (Coleoptera, Ptiliidae). *Zool. Zh.* **84**, 181–189 (in Russian).

Polilov, A. A. (2008). Anatomy of the smallest Coleoptera, featherwing beetles of the tribe Nanosellini (Coleoptera, Ptiliidae), and limits of insect miniaturization. *Entomol. Rev.* **88**, 26–33.

Polilov, A. A. and Beutel, R. G. (2010). Developmental stages of the hooded beetle *Sericoderus lateralis* (Coleoptera: Corylophidae) with comments on the phylogenetic position and effects of miniaturization. *Arth. Struct. Dev.* **39**, 52–69.

Polivov, A. A. and Beutel, R. G. (2009). Miniaturization effects in larvae and adults of *Mikado* sp. (Coleoptera: Ptiliidae), one of the smallest free-living insects. *Arthrop. Struct. Dev.* **38**, 247–270.

Quesada, R., Triana, E., Vargas, G., Seid, M., Douglass, J., Eberhard, W. G. and Wcislo, W. (in press). Large brains in small spiders: the central nervous system extends into the legs of miniature spiders. *Arthr. Struct. Develop.* doi:10.1016/j.asd.2011.07.002.

Reed, C. F., Witt, P. N. and Jones, R. L. (1965). The measuring function of the first legs of *Araneus diadematus* Cl. *Behaviour* **25**, 98–118.

Reed, C. F., Witt, P. N. and Scarboro, M. B. (1969). The orb web during the life of Argiope aurantia (Lucas). *Dev. Psychobiol.* **2**, 120–129.

Reed, C. F., Witt, P. N., Scarboro, M. B. and Peakall, D. B. (1970). Experience and the orb web. *Dev. Psychobiol.* **3**, 251–265.

Rehen, S. K., McConnell, J. J., Kaushal, D., Kingsbury, M. A., Yang, A. H. and Chun, J. (2001). Chromosomal variation in neurons of the developing and adult mammalian nervous system. *Proc. Natl. Acad. Sci. USA* **98**, 13361–13366.

Rensch, B. (1948). Histological change with evolutionary changes of body size. *Evolution* **2**, 218–230.
Rensch, B. (1960). Evolution above the Species Level. Columbia University Press, New York, NY.
Riska, B. and Atchley, W. (1985). Genetics of growth predict patterns of brain-size evolution. *Science* **229**, 668–671.
Riveros, A. J., Seid, M. and Wcislo, W. T. (submitted). Evolution of brain size in class-based societies of fungus-growing ants. *Animal Behaviour*.
Roth, L. M. and Willis, E. R. (1958). The reproduction of cockroaches. *Smithsonian Misc. Coll.* **122**, 1–49.
Roth, G., Rottluff, B. and Linke, R. (1988). Miniaturization, genome size and the origin of functional constraints in the visual system of salamanders. *Naturwissenschaften* **75**, 297–304.
Roth, G., Rottluff, B., Grunwald, W., Hanken, J. and Linke, R. (1990). Miniaturization in plethodontid salamanders (Caudata: Plethodontidae) and its consequences for the brain and visual system. *Biol. J. Linn. Soc.* **40**, 165–190.
Roth, G., Blanke, J. and Wake, D. B. (1994). Cell size predicts morphological complexity in the brains of frogs and salamanders. *Proc. Natl. Acad. Sci. USA* **91**, 4796–4800.
Rutowski, R. L. (2000). Variation of eye size in butterflies: inter- and intraspecific patterns. *J. Zool. Lond* **252**, 187–195.
Sandoval, C. P. (1994). Plasticity in web design in the spider *Parawixia bistriata*: a response to variable prey type. *Funct. Ecol.* **8**, 701–707.
Sarko, D. K., Catania, K. C., Leitch, D. B., Kaas, J. H. and Herculano-Houzel, S. (2009). Cellular scaling rules of insectivore brains. *Frontiers in Neuroanatomy* **3**, 1–12.
Sarpeshkar, R. (1998). Analog versus digital: Extrapolating from electronics to neurobiology. *Neural Computation* **10**, 1601e1638.
Scharff, N. and Coddington, J. A. (1997). A phylogenetic analysis of the orb-weaving spider family Araneidae (Arachnida, Araneae). *Zool. J. Linn. Soc.* **120**, 355–434.
Schoenemann, P. T. (2004). Brain size, scaling and body composition in mammals. *Brain Behav. Evol.* **63**, 47–60.
Seeley, T. D. (1985). Honey Bee Ecology. Princeton University Press, Princeton.
Seeley, T. D. (2010). Honey Bee Democracy. Princeton University Press, Princeton.
Seid, M. A., Castillo, A. and Wcislo, W. T. (2011). The allometry of brain miniaturization in ants. *Brain Behav. Evol.* **77**, 5–13.
Sherry, D. F., Forbes, M. R. L., Khurgel, M. and Ivy, G. O. (1993). Females have a larger hippocampus than males in the brood-parasitic brown-headed cowbird. *Proc. Natl. Acad. Sci. USA* **90**, 7839–7843.
Shettleworth, S. (2010). Cognition, Evolution, and Behavior. 2nd ed. Oxford University Press, New York, NY.
Smith, D., Wessnitzer, J. and Webb, B. (2008). A model of association learning in the mushroom body. *Biol. Cybern.* **99**, 89–103.
Spaethe, J. and Chittka, L. (2003). Interindividual variation of eye optics and single object resolution in bumblebees. *J. Expt. Biol.* **206**, 3447–3453.
Stachowitsch, M. (1992). The Invertebrates. Wiley-Liss, New York, NY.
Stevens, S. S. (1975). Psychophysics. John Wiley & Sons, New York, NY.
Stewart, P. (1964). Jacobellis v. Ohio. (U.S. Supreme Court 378 U.S. 184).
Strathmann, R. R. (1975). Larval feeding in echinoderms. *Am. Zool.* **15**, 717–730.
Strausfeld, N. (1976). Atlas of an Insect Brain. Springer, Berlin.
Striedter, G. F. (2005). The Principles of Brain Evolution. Sinaur Assoc. Inc., Sunderland, MA.
Tinbergen, N. (1951). The Study of Instinct. Oxford University Press, Oxford.

Tononi, G., Sporns, O. and Edelman, G. M. (1994). A measure for brain complexity: relating functional segregation and integration in the nervous system. *Proc. Natl. Acad. Sci. USA* **91**, 5033–5037.

Van Essen, D. C. and Gallant, J. (1994). Neural mechanisms of form and motion processing in the primate visual system. *Neuron* **13**, 1–10.

van Wijk, M., Wadman, W. J. and Sabelis, M. W. (2006). Gross morphology of the central nervous system of a phytoseiid mite. *Expt. Appl. Acarol.* **40**, 205–216.

Vollrath, F. (1987). Spiders with regenerated legs can build normal webs. *Nature (London)* **328**, 247–248.

Vollrath, F. (1992). Analysis and interpretation of orb spider exploration and web-building behavior. *Adv. Stud. Behav.* **21**, 147–199.

von Frisch, K. (1967). The Dance Language and Orientation of Bees. Harvard University Press, Cambridge.

Vos, M., Hemerik, L. and Vet, L. E. M. (1998). Patch exploitation by the parasitoids *Cotesia rubecula* and *Cotesia glomerulata* in multi-patch environments with different host distributions. *J. Anim. Ecol.* **67**, 774–783.

Walter, D. E. and Proctor, H. C. (1999). Mites, Ecology, Evolution and Behavior. CABI Publishing, New York, NY.

Warrant, E. J. (2008). Seeing in the dark: vision and visual behaviour in nocturnal bees and wasps. *J. Exp. Biol.* **211**, 1737–1746.

Warrant, E. J., Kelber, A., Gislén, A., Greiner, B., Ribi, W. and Wcislo, W. T. (2004). Nocturnal vision and landmark orientation in a tropical halictid bee. *Curr. Biol.* **14**, 1309–1318.

Watts, D. J. and Strogatz, S. H. (1998). Collective dynamics of 'small world' networks. *Nature* **393**, 440–442.

Wcislo, W. T. (1995). Sensilla numbers and antennal morphology of bees (Hymenoptera: Apoidea), in relation to behavior and taxonomy. *Int. J. Insect. Morphol. Embryol.* **24**, 63–81.

Wcislo, W. T. and Tierney, S. (2009). Behavioral environments and niche construction: the evolution of dim-light foraging in bees. *Biol. Rev. Cambr. Phil. Soc.* **84**, 19–37.

Wehner, R. (1987). 'Matched filters' - neural models of the external world. *J. Comp. Physiol. A* **161**, 511–531.

Wehner, R., Fukushi, T. and Isler, K. (2007). On being small: brain allometry in ants. *Brain Behav. Evol.* **69**, 220–228.

Wheeler, W. C. and Hayashi, C. Y. (1998). The phylogeny of the extant chelicerate orders. *Cladistics* **14**, 173–192.

White, J. G., Southgate, E., Thomson, J. N. and Brenner, S. (1975). The structure of the ventral nerve cord of *Caenorhabditis elegans*. *Philos. Trans. R. Soc. Lond. B* **275**, 327–348.

Wille, A. (1961). Evolutionary trends in the ventral nerve cord of the stingless bees (Meliponini). *Rev. Biol. Trop.* **9**, 117–129.

Wilson, E. O. (1984). The relation between caste ratios and division of labor in the ant genus *Pheidole* (Hymenoptera: Formicidae). *Behav. Ecol. Sociobiol.* **16**, 89–98.

Windhoffer, R. and Westheide, W. (1988). The nervous system of the male *Dinophilus gyrociliatus* (Annelida: Polychaeta). I. Number, types and distribution pattern of sensory cells. *Acta Zool. (Stockholm)* **69**, 55–64.

Witt, P. N. (1965). Do we live in the best of all possible worlds? Spider webs suggest an answer. *Perspect. Biol. Med.* **8**, 475–487.

Witt, P. N., Reed, C. F. and Peakall, D. B. (1968). A Spider's Web. Problems in Regulatory Biology, Springer-Verlag, New York.

Wittenberg, G. M. and Wang, S. S. H. (2007). Evolution and scaling of dendrites. In: *Dendrites* (eds Häusser, M., Spruston, N. and Stuart, G.), 2nd ed., pp. 43–67. Oxford University Press, New York, NY.

Yurov, Y. B., Iourov, I. Y., Monakhov, V. V., Soloviev, I. V., Vostrikov, V. M. and Vorsanova, S. G. (2005). The variation of aneuploidy frequency in the developing and adult human brain revealed by an interphase FISH study. *J. Histochem. Cytochem.* **53**, 385–390.

Combined Index of Volumes 40 and 41

The numbers 'V40' and 'V41' following locators refers to volumes 40 and 41 respectively.

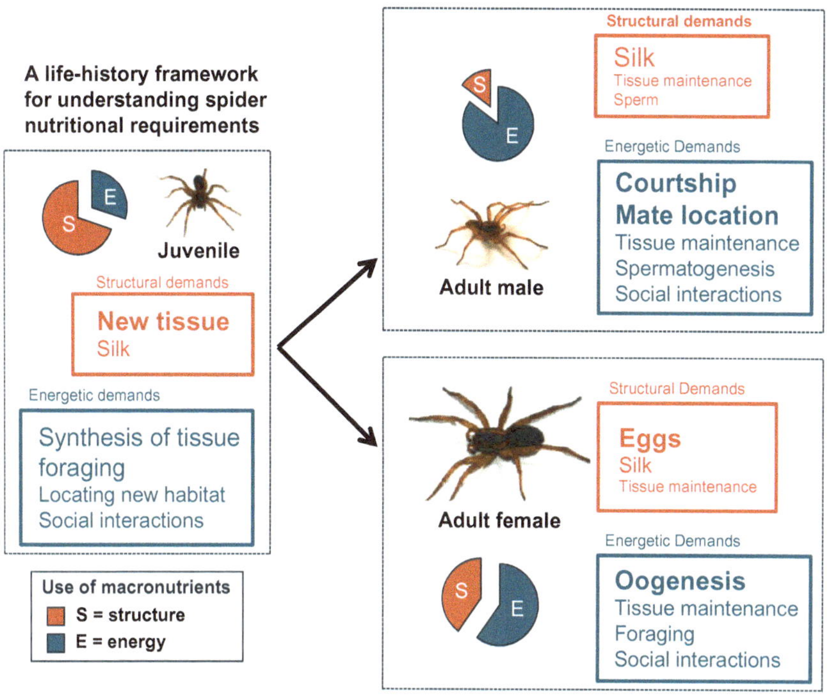

PLATE 1 Hypothesized relationships between life history characteristics and the nutritional requirements of spiders. Major energetic and nutritional demands of the different sexes and life stages are listed with font sizes proportional to the relative requirements of those activities. "Structural demands" refer to the use of nutrients to synthesize new molecules in the body, while "energetic demands" refer to the catabolism of nutrients to fuel metabolic activities. (For b/w version, see page 123 in the volume.)

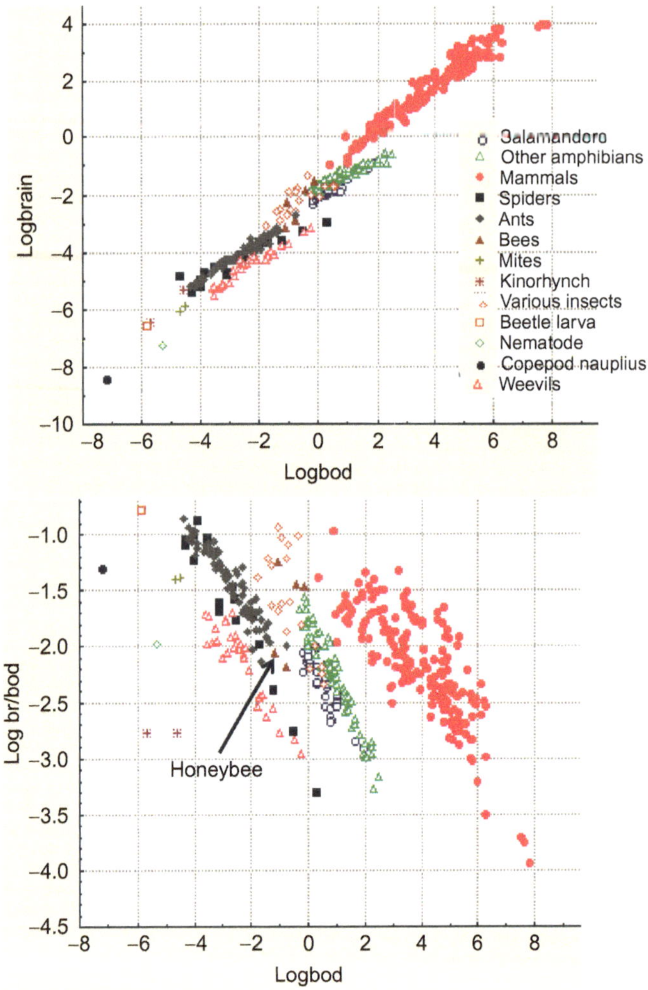

PLATE 2 Brain weight and body weight in vertebrates and assorted invertebrates are given in log–log plots of brain weight versus body weight (A) and of the fraction of the body weight devoted to the brain versus body weight (B) (which permits clearer contrasts between different groups). Sources of data are the following. Vertebrate data were read from graphs in Striedter (2005). Bee weights are from Mares *et al.* (2005), ant weights are from Wehner *et al.* (2007) and Seid *et al.* (2011); weevil weights are from M. Seid, H. Stockwell and W. Wcislo (in preparation). Spider and mite weights were measured directly and from reconstructions of brains from serial sections (Quesada *et al.*, in review, G. Vargas, unpublished). A drawing of a ''typical'' kinorhynch body and brain (Barnes *et al.*, 1993) was deconstructed into simple geometric forms and volumes were estimated based on the cross-sectional body outline in Fig. 296 of Higgins (1983). Two points are given, reflecting uncertainty regarding body weight. Estimated nervous system volume may be underestimated more seriously in this species due to its having a relatively large nerve cord (Bullock and Horridge, 1965). Similar deconstructions and analyses utilized a

drawing of the anterior end of the nematode *Caenorhabditis elegans* and a photograph of its entire body (by White *et al.*, 1975), and a SEM photo of the entire body plus a cross-section of the body of the first nauplius larva of the copepod *Dactylopusia* sp. (Lacalli, 2009, Figs. 1A and 1D) (this volume may be an underestimate, as the bulbous outline of the nauplius was not obvious in the SEM photo—T. Lacalli, personal communication). The brain volume of this nauplius was calculated by reconstructing sections of the neuropil examined under the TEM (Lacalli, 2009), and an estimate that the cell bodies occupy a volume a bit larger than that of the neuropil (T. Lacalli, personal communication). The body volume of the beetle larva *Mikado* sp. was estimated by deconstruction from Fig. 5 of Polivov and Beutel (2009), assuming a cylindrical cross-section. Their graph (Fig. 20) indicates that the brain occupied about 16.5% of the body volume, so this value was used to estimate the brain volume. All calculations assumed that one cc of brain tissue weighs one gram. All points, other than the copepod nauplius, the beetle larva, and four spiders refer to adults (thus for most groups, additional values for smaller body sizes are lacking). The "brains" of spiders and mites include tissue that in a vertebrate or an insect would be included in the spinal cord or ventral nerve cord, and are thus somewhat higher than they should be to make proper comparisons (for instance, the allometric line for spiders in (b) should probably be below that for ants). (For b/w version, see page 159 in the volume.)

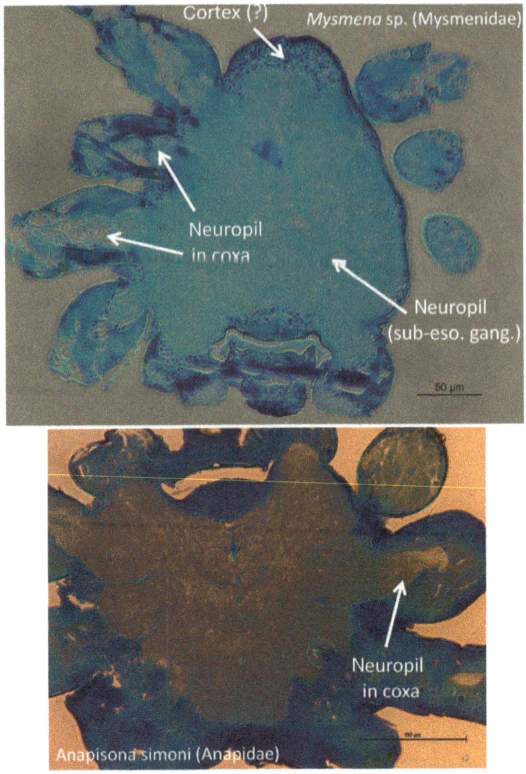

PLATE 3 Sections of tiny spiders showing extensions of the brain neuropil into their coxae. (For b/w version, see page 172 in the volume.)

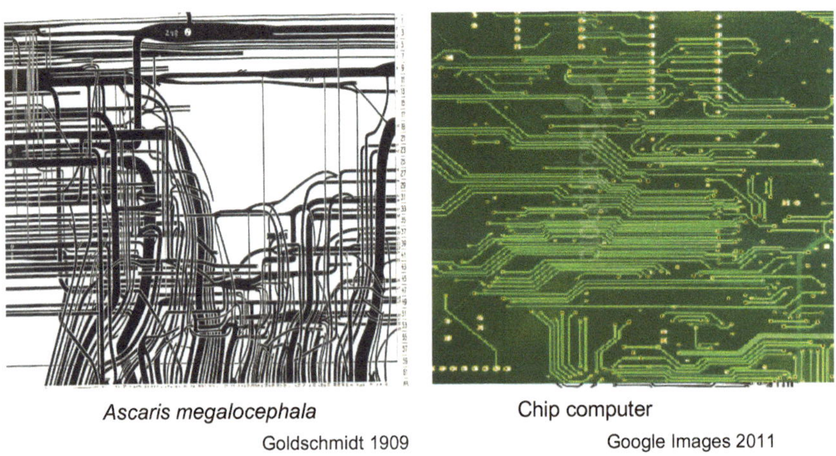

PLATE 4 Diagrammatic representation (left) of the connections of 70 of the 134 fibres in the circumenteric ring near the origin of the ventral nerve cord of the nematode *Ascaris megalocephala* (from Bullock and Horridge, 1965, after Goldschmidt, 1909). The uncanny visual resemblance to a computer chip (right) emphasizes the computer-like traits of nematode nervous systems (fixed numbers of elements, invariant connections between them) that are exactly the computer-like aspects that do *not* occur in vertebrate nervous systems (Edelman, 1998). This emphasizes the likely profound differences in how the brains of taxa in different grades function. (For b/w version, see page 204 in the volume.)

Kolibri 360 GmbH
Ferdinand-Jühlke-Straße 7
99095 Erfurt, Deutschland
produktsicherheit@kolibri360.de